MW00453940

The
Visual
Detox

The Visual Detox

HOW TO CONSUME MEDIA WITHOUT LETTING IT CONSUME YOU

Marine Tanguy

1 3 5 7 9 10 8 6 4 2

Square Peg, an imprint of Vintage, is part of the Penguin Random House group of companies whose addresses can be found at global.penguinrandomhouse.com

Penguin
Random House
UK

First published by Square Peg in 2024

penguin.co.uk/vintage

Typeset in 12.15/14.75pt Dante MT Std by Jouve (UK), Milton Keynes
Printed and bound in Great Britain by Clays Ltd, Elcograf S.p.A.

The authorised representative in the EEA is Penguin Random House Ireland, Morrison Chambers, 32 Nassau Street, Dublin D02 YH68

A CIP catalogue record for this book is available from the British Library

ISBN 9781529912647

Penguin Random House is committed to a sustainable future for our business, our readers and our planet. This book is made from Forest Stewardship Council® certified paper.

MIX
Paper | Supporting
responsible forestry
FSC® C018179

To my boys Atlas and Vivaldi, Maman is so excited to
see what visual narratives you will create and
I will be here to support them.

To my granny who taught me about being resilient, Louise
who taught me about friendship, William who taught me
about love, Lise who taught me how to be a better CEO,
Jean Luc who taught me about family love, and David
whose talent has changed my life. I still have plenty
to learn from all of you.

To my mother Isabelle, my father Fabrice and my sister
Sarah; for everything we cannot say and yet see.
There is always the sea.

Onwards X

Contents

Contents

PART II

Your Visual Detox

Contents

Introduction

By the time you reach the end of this sentence, 1,250,000 new images will have been uploaded online.[1] That's about 1.8 billion new visuals per day. To put this into perspective, humans with sight can process about 36,000 visual messages per hour, or ten images per second, which we absorb and process regardless of whether we are consciously thinking of them. How often do you reach for your phone just for a quick distraction? We live in an age of visual abundance and virtually anything you want to see is always at your fingertips – whether it's a friend's new holiday photo, a funny cat video or the world's most famous and moving works of art. When we venture online, we witness an estimated average of up to 10,000 commercial visuals a day. These visuals, ranging from static images to video clips and animations, are constantly available to us and vying for our attention, so it's no wonder that we sometimes feel stressed, tired or anxious as a result – our brains are regularly working in overdrive to process everything that we see.

Whether this leaves you feeling intrigued or a little spooked, the truth is that what we see is a fundamental part of our daily lived experience. Whether we witness, feature, upload or share them we often have no choice but to consume images, but this consumption does not have to be overwhelming and draining. Visuals can also have a powerful and affirming impact on our lives. Our minds often work in pictures and images rather than words: We are largely visual learners – 65% of what we learn is through sight[2] and we remember 65% of what we see, compared to only 10% of what we hear.[3] Visuals can help us communicate efficiently, see things from a new perspective and even tap into different emotional states. Whether we are online or not, we are constantly seeing images, and over time this daily consumption shapes our visual narrative – the story we tell ourselves about who we are and the world around us based on the visual information we've absorbed. Over time, this daily consumption becomes our visual narrative – the story we tell ourselves about our lives and the world around us created by that visual consumption.

The key to tapping into the benefits of our world's visual abundance to improve our overall wellbeing is to decode how our visual consumption is affecting our life and take back control of our visual narrative. As the founder of the leading talent agency in the UK art sector, I have devoted my life to curating visuals that powerfully impact people's lives, understanding how they can be used to shape how we think and feel and making the positive effects of art available to everyone. Through *The Visual Detox* I will give you the tools that I use in my own life to focus on what is important and meaningful, resist visual manipulation and make your visual landscape inspiring and empowering.

Visual Manipulation

Every image that we see is trying to tell us something, and a substantial number of them are trying to sell us something, or at least sell us on an idea. We have a tacit understanding that witnessing a barrage of words and, in particular, foul language can feel overwhelming or insulting, but when it comes to what we see, we rarely regard intrusive images with the same disquiet. Advertisements follow us throughout our day, across every corner of the internet and in our offline lives, on billboards, shop windows and even painted on to passing vehicles. What does such a deluge of visuals do to our wellbeing? We are constantly being told what to do and who to be through visuals, and these images add up, sinking into our subconscious and influencing our thoughts and choices.

Unless you are already actively curating your visual intake, whenever you look at a screen, billboard or other media portal, you are likely to be encountering the following three messages:

1. Be sexy – this is what sexy is . . .
2. Be intimidated – you are not as _____ as this . . . [please fill in the blank]
3. Believe this, not that . . .

If these were literal words, you'd probably find yourself swearing at ads a few times a day. But most of us don't do so because we are not even aware of the messaging. And that's the 'soft' power of visuals – they make us do, think and feel things that we would never do, think or feel if that message was explicitly spelt out in words. When we consume images without understanding that the message they are sending us is having an effect on our thoughts and wellbeing, we become indifferent to the message and continue to consume these visuals without interrogating them.

The biggest dangers that come from not curating your visual consumption are poor mental health, negative body image, overconsumption and political manipulation. Indifference to what visual messages we consume leaves us open to manipulation, but it is possible for everyone to learn how to feel connected to the visual language that is being used to speak to us every day. This extends to understanding the processes that determine the curation of what we see in the first place. Art is not just a portrait on a gallery wall, it is the human application and expression of creativity to produce objects, environments and experiences to communicate with each other and enhance our lives. It affects all of us. Our visual consumption is instrumental in taking charge of that process for ourselves to create a supportive and nourishing visual narrative.

I started my career as a gallerist, where I spent my days watching how people reacted to art. The right image could bring a person to tears or revitalise and inspire them. Recognising this power that visuals have over how we think and feel, I realised that the traditional models for art in society were not doing enough to benefit people from all walks of life, so I created MTArt, a talent agency that is structured as a B Corp with public interests at heart. The art world isn't just about the buying and selling of artworks, it is also about creating the self-image of entire nations, transforming neighbourhoods and cities, and the fashioning of social and diplomatic identities for groups within and beyond national boundaries. Our mission is to widen audiences for art and go beyond the traditional buying, selling and commissioning of artworks by creating public art projects, digital projects and collaborative projects that enable our artists to bring their work into the heart of the society we live in, where they can be part of our shared visual landscape. In this way our visual surroundings (streets, squares, metaverse and homes) can be transformed with the art of the most talented artists of our generation, reaching people from all walks of life.[4]

Over the course of this book, I will teach you how to understand the visual messages you are consuming and what you can do to change the visual landscape that shapes your life. In the first part of the book, we will focus on improving your visual literacy by exploring how visuals impact us, how the context in which we see them shapes their impact and meaning, and how you can learn to 'read' an image so that you consciously understand what your unconscious mind is absorbing. Along the way, I will give you the tools to perform a visual audit to better understand the visual messages you're receiving day in and day out. In the second part, we will focus on how other people decide what we see (and what we don't) and what you can do to become your own curator while detoxing from the parts of your visual consumption that are not serving you. Throughout the book I've brought together research that we've done at MTArt on the social impact of public art, and case studies from artists like Sara Shakeel and Adam Nathaniel Furman about how their visual consumption shapes their life, and from ordinary people like journalist Yalda Hakim and Mark Maclaine, a British educator, music producer and founder of Tutorfair, who have felt the effects of their visual consumption and taken steps to detox. I have also provided multi-disciplinary research from around the world on how we process visual messages, how messages are curated for us and the psychological and sociological impacts of common visual consumption trends. Then, we'll take a step back and explore how we can look at our cities and community spaces in a new light to transform the visuals that shape our physical surroundings.

My aim with *The Visual Detox* is to start a movement – where people are empowered and develop their visual literacy skills. As a woman, I wanted women to feel represented, supported and empowered visually. As an art lover, I wanted the arts to represent all artistic voices, not just the very few artists and art collectors who can afford to be in the art world. As the mother of two boys, I want them to address the world visually. I hope this book starts a stronger relationship with the visual world that you take part in every day. Welcome to the language of visuals.

After all that, I too have an idea I want to sell you on: the idea that it doesn't need to be this way. We don't need to be overstimulated, patronised, and manipulated by visuals. We can lead the visual world, contribute to it and make this language represent all of us. Technology and media should inform and help us, not oppress us. If you agree, stay with me.

To begin, let's start with an experiment.

Part I

Your Visual World

Chapter 1.

Understanding Your Visual Consumption

What is Your (Visual) Type?

Where does your mind go to when you read the word 'apple'? What do you see in your mind when you close your eyes? Do you just see five letters? A 2-D image of an apple? Or can you visualise and perhaps even taste a crisp and juicy fruit? Try it and see.

By completing this exercise, you have just taken part in the infamous apple test described by Firefox founder Blake Ross in his viral Facebook post on discovering that he has aphantasia, a rare condition that causes people to be unable to see in their imagination.[1] When they close their eyes, they don't see anything at all. This may seem like a minor oddity but mental imagery is in fact an important component of other brain functions, particularly of memory processes.[2]

Most likely, you did see something when you pictured an apple. Most of us do and we frequently think in pictures. For about 60–65% of humans this is the primary way we experience our thoughts. Research has shown that we can identify distinct groups of visual thinkers based on their preferred mode of visual processing.[3] There are visual thinkers, spatial thinkers and verbal thinkers – how each one prefers to visualise concepts ends up being predominantly how they learn.[4] Many people are able to tap into more than one kind of visual thinking, but we all have a preference. Knowing what our type is can help open up new avenues for learning and make us more aware of our blindspots.

Visual Thinkers

For the 60–65% of us who are visual thinkers, we predominantly understand and learn things through picturing them in our minds. Many artists fall into this category. Visual thinkers are good at taking in visual information and understanding what they're seeing, and creating new images within their minds. When they picture an apple, they can easily visualise a wide range of apples, including different varieties or illustrations of apples. They often learn new concepts better when they can see them illustrated and have an aptitude for recognising patterns, creating visuals that draw people in. Temple Grandin, American scientist, autism advocate and animal behaviourist known for her work in understanding visual thinking, describes visual thinkers as having a superpower when it comes to non-linear thinking: they are great at making quick associations between seemingly disparate bits of information and picking up on details that other people miss.[5]

Spatial Thinkers

Closely related to visual thinking is spatial thinking. Temple Grandin includes spatial thinking within her explanation of visual thinking, but I think it is useful to think of spatial thinkers within their own category because spatial thinking is more focused on looking at patterns and relationships within 3-D space. Spatial thinkers also rely on largely visual ways of thinking but they have a particular preference and aptitude for understanding relationships between objects in space. For example, when they visualise an apple, they likely picture a 3-D apple that they can rotate within their mind's eye. They're usually good at navigating, understanding complex machinery and problem solving that involves making physical objects fit together. Spatial thinkers are often found among our engineers, architects and designers of 3-D spaces. Their grasp of dimensionality is how they see and think and create, spatially and visually, when they look at things. Schematics such as blueprints speak volumes of information to them that perhaps escapes the visual or verbal thinker.

Verbal Thinkers

Verbal thinkers, on the other hand, are focused primarily on words. They will often think through things using an inner monologue or by writing them down and reading them back. When they close their eyes and think about 'apple', they often see the word written out or hear the word in the languages they know. They are predisposed to thinking through things in a linear way, which often makes them good at solving abstract problems that don't have a visual component and presenting information to others.[6]

Case Study

*How my visual thinking
shaped MTArt*

I am most definitely a visual thinker. I imagine every-thing as a picture – visually – before putting it into words. I see the colours, the textures of my ideas, before I have the words to communicate them. At first it was a problem: how do you translate such a strong visual idea into words? My French literature professor used to say to me that an idea doesn't exist without words, but I happily disagree today. An idea can start visually. When I first set up MTArt, I imagined my entire company visually, from the kind of projects we'd achieve, to the office itself, from the receptionist at the door, all the way down to the type of desks people would sit at. This vision of what I wanted the company to look like was my guide as I made decisions that would get me closer to fulfilling my vision. My brain comprehends things better and feels more confident when I see things. This is why, instead of looking for a business partner who would put some cash in, or some investors, I wanted to witness the first stages of the company and have a direct role in all the decisions needed to set it up and make it profitable before scaling it up. I needed to see every layer as it scaled, from the foundations upwards, to deeply understand how everything came together to move forward with the confidence I needed.

For example, my first office was in the derelict former hospital space owned by Camden Collective – a charity for creative spaces that offers co-working space for creatives alongside business space for some incredible entrepreneurs like Anne Marie Imafidon who co-founded the Stemettes, a social enterprise group which supports girls, young women and non-binary young people in STEM.[7] I loved this space – it really appealed to my visual thinking brain. My life was a shambles: I was broke, struggling, and in an office filled with whiteboards and Post-it notes, with a horrible purple carpet. I was wearing trainers everywhere and cycling around London to studios to make our first deals and sign our first artists. However, my visual narrative felt authentic and full of potential. Everywhere I looked, I saw a reflection of where I was on my journey, the progress I was making day by day and physical representations of my plans for the future that inspired me to keep going. The office was run down at first, but that room for improvement is what gave me the space to visualise what it could become. The co-working space of that hospital provided me with a cohort of really great creative and impactful entrepreneurs and a bare-bones visual environment as a jumping-off place. That's how the building played a role in our evolution: it could serve as a blank canvas for our vision that liberated us from the fear of messing up something too perfect. We had the freedom to transform and reinvent our space, which we did: we covered the front of that hospital for a festival; we did performances on the streets around the building, painted the façade of the Camden Collective and reinvented offices around us. In this way, within only a few months we started to reinvent the world visually around us with our clients, spreading out into the surrounding community.

Having this evolving base to work from that reflected how we were progressing inspired us to work hard and reach further: by the end of our first year we were profitable, striving and excited to take on the world. Eventually, my reality started to look like the vision that I had starting out. For me, that's the superpower of being a visual thinker: you just start with a given or provided reality and you transform it through personal creativity by taking control, bit by bit, of the visuals around you and changing them to align with the best version of reality that you can imagine.

Exercise
Discover Your Visual Type

I find that a good way to identify what kind of thinker you are is to go to a museum and see what your attention is drawn to:

- If you, like 35% of the population, find yourself drawn to the little text next to an artwork at the museum and prefer to learn about the piece before looking at it deeply, you are more likely a **verbal thinker**. When you look at a piece of art, you feel drawn to knowing about the artist or the context behind the piece.

- If you find your attention drawn to the shape of the exhibition space, how the piece sits within that space, and the composition, structure and lines within the visual first, and your initial questions about a piece of art often sound like 'How was this built?' or 'How do these different parts work together?', then you are most likely a **spatial thinker**.

- If your first response to a piece of art is to be attracted to the colours, the feelings of the visuals and the instinctive connection you have with it, you may be a **visual thinker**. Your first thoughts looking at a piece of art often include imagining it in other spaces to see how it would look and thinking about what other visuals it reminds you of.

The key takeaway is that most of us experience our thoughts primarily through visuals, evidence that visual codes are deeply embedded in our lived experience and that this is a two-way street: what you see triggers thoughts, and your thoughts often appear to you in visual form. This gives you the opportunity to use visuals to shape your mind and use your mind to visualise what you wish to see in the world. Now that you know the kind of thinker you are, you can choose to prioritise content that speaks to your preferred ways of processing information when you need to learn something new. No matter what your visual type is, the next chapters will help you consciously process the messages we absorb from visuals so that you can understand how visuals are affecting you.

My Visual Narrative

There is an adage that says we are defined by the five people we spend the most time with. The same conclusion might be drawn with visual consumption. We are in part defined by the visual environments we spend the most time in. Your first step towards visual literacy is to start reflecting on the visuals in your daily life, and to recognise the feelings they evoke in you – whether good or bad.

For example, the first year of my life was largely spent in darkness. I don't remember this consciously, but I was told as much by my grandparents and my mother. The shutters on our windows were always closed. Mum was depressed, initially because Dad had left her when she was pregnant with me (he came back a year later) and then, once I was born, due to severe postnatal depression. We lived in a small rural community in the eighties so most people – including my mum – didn't know what postnatal depression was, never mind how to treat it. We basically hid away in a tiny room of our house for the first year and she did her very best to get through each day. In those formative moments, my home life was bleak. I would describe my early childhood as a series of greys, blacks and deep blues.

In his brilliant book, *The Body Keeps the Score*, author Bessel van der Kolk demonstrates that children who experience trauma are slower to speak.[8] I was quiet until I was seven years old and so my world was mostly visual as opposed to verbal. I didn't realise it until much later, but this made me visually sensitive.

Van der Kolk's book taught me that this is common for kids who are in an environment that doesn't feel fully emotionally secure. Such children tend to internalise more and observe outwardly more, because their security (physical, mental, emotional) is at stake. They come to understand that they can't rely on parents and become hyper vigilant as their own protector. They watch out for quickly changing situations, as they might know from experience at a very young age that at one point you could be happy and feeling safe, at another point you could be suddenly super unhappy or feeling insecure. I understood unconsciously that the bare and sterile environment I encountered at home was linked to my parents' emotional state.

As I got older, I was left to my own devices and would disappear from the house at every opportunity. My hometown, Île de Ré, is a beautiful little French island, home to around 9,000 people when I was born, a nature reserve and bird sanctuary. It is a visual sanctuary. I often spent all day cycling through the countryside or along the coastline – anything to escape the hostility and sadness at home. This environment made me feel awe at how beautiful the natural world can be and I was grateful for how lucky I was to exist within it.

While those outdoor environments were a space of quiet beauty and calm, my life at home was still tense. My mother's poor mental health meant her escape was to compulsively clean all day. My parents both valued practicality – they didn't keep objects around for aesthetic or sentimental reasons – everything had to serve a purpose. The bare interior we lived in up until I was seventeen years old was spotlessly clean but devoid of life with its barren whites, faded greys and greens. When I would come home and see this space it would send the message that life, expression, joy and feelings had no place in such a spotless, sterile-looking environment. It is still a vivid memory. It makes me feel anxious just thinking about it. Today, I have so much joy looking at the chocolate ice cream dripping down the shirt of my eldest son as he smiles. This mix of colours and lack of perfection is my daily reminder that joy and emotions won over perfection.

Although I feel like I was born with an intuitive understanding of the nature and power of visuals, I have also chosen to purposefully craft my visual narrative and hone my visual literacy. I studied Art History at the University of Warwick and have spent my career focused on understanding the power of visuals and how they can be used to reshape our lives, but you don't need a formal art degree to feel how your visual consumption affects your life.

Case Study
Sara Shakeel

Pakistan, 1996. From the heart of the bustling city of Islamabad, artist Sara Shakeel's earliest memories of home can be described in a constellation of colours. The compound in which she resided was expansive, with light filling most of the space. Windows reached from floor to ceiling and opened out onto luscious expanses of flora and fauna, in hues of emerald, ochre and amber. The interior was mainly painted white, yet as she describes, there were hints of earth tones throughout the house, such as furniture made of walnut, cedar and rich oak. They sat against colours of intense grandeur, golds and platinums, which lined various parts of the home.

Shakeel's mother was the principal of a school situated in the compound Shakeel grew up in, which allowed her the opportunity to travel to the school library outside of opening hours. There, she browsed through the variety of books, taking in the rich scents and textures of paper, old and new. Her father was also a teacher, and was constantly engrossed in some sort of handiwork. 'If something broke, he would fix it, and then decorate it with hundreds of beads,' Shakeel tells me on a sunny day in my office. The family home possessed numerous crystal chandeliers affixed to the ceiling, and when one chandelier broke, her father decided to lay each piece out on the dining table to fix it. There, the artist witnessed dozens of shards of crystal sprawled out, producing a kaleidoscope of refractions. Shakeel was amazed by their duality – how they appeared 'clear from the outside' yet looking through them 'showed every colour of the rainbow' – for her it was a chromatic explosion.

Inside the confines of her home, she would describe colours as containing a sort of privilege she was aware of. But out on the street, the artist witnessed immense poverty, which triggered in her great sadness and melancholia, in hues of mauve, coal and onyx. This contrast of life in Pakistan, between rich and poor, light and the dark, joy and sorrow, was something Shakeel has become aware of in other cities of the world, such as New York and also London (where she currently resides).

Those disparities became her visual narrative and today, Shakeel has become one of the most followed artists on Instagram (over 1 million followers) and been able to connect with people from across the world. In her art there are a variety of objects which could be described as banal: a peeled orange; a young cat; an aeroplane stationed on the tarmac; even her *Glitter Stretch Marks* is a photographic portrait in browns of pregnancy, banal at the same time it is striking. Yet, they are contrasted with an abundance of crystals, which consume most of the composition. It is this very aesthetic, which mimics the kind of visual inequalities the artist witnessed in her youth – the mixture of something described as average and ordinary, juxtaposed with the grandiose or divine – that lies at the heart of her compositions.

Our visual landscapes don't just affect our personal lives, they also shape our visual narratives in our professional lives. In my early twenties I got to know Andrew Lamberty, a successful gallery owner who split his time between London, New York and Italy. When he was invited to show his gallery at the prestigious Pavilion des Arts et du Design, PAD, he hired me to help run the gallery's stall. He believed that only wealthy people succeed in the art sector and that women in the art world needed to look the part. He therefore commented on how I could improve my outfits to adapt my visual appearance to the upper classes, which led me to try and learn the visual codes to succeed in this task. Little wonder that his influence started to alter not only my visual consumption but the visual narrative that I constructed as a result. I transformed myself visually. Inspired by how Renaissance painters would communicate wealth and status in their subjects by including valuable and symbolic items in their portraits, I started to wear large hats and pose with glasses of champagne in all of my photographs and collected objects that signified wealth and status, making sure to include them in photos and conversations. The environment I was in made me sensitive to the visual language of class and wealth and comparing myself to those images changed how I lived my life.

As a result of my work at PAD, I was then approached to be Gallery Manager of the Outsiders Gallery in Soho, London. The title sounded prestigious, but the pay was low and while I was socialising with the wealthy London elite, I was living in a shared room and sleeping on a sofa bed. My visual narrative could not have looked further from the image I was projecting, but I was building up a reputation in the art world. After I started work there, advertising investor Steph Sebbag approached me to start my own gallery in Los Angeles. We became business partners and I went to Los Angeles with him to open the De Re Gallery, named after the island I come from, Île de Ré.

When I moved to Los Angeles in 2014, I once again experienced how much the visual landscape you spend time in shapes your life. When I lived in LA, I would walk to work – a Herculean task for the pedestrian-shy city – and find myself surrounded by adverts and images that reinforced what a woman should look like in Los Angeles. The gallery was in the famous Beverly Hills, and my home was in the more residential Hollywood Hills; both expensive locations were sponsored by my first investor. Each day I'd walk from my home to the gallery on Melrose Avenue, one of the biggest shopping streets in the area, and, all along my route, consume hundreds of those images shaping what my narrative needed to be. When you walk this route, you encounter incredible views of luxury homes, swimming pools and tennis courts with green foliage, despite the dry hot weather. This is one visual aspect of the American dream: the architecture of the houses, the gardens, the tennis courts, the swimming pools are also very imposing and the view from the street serves as a visual display of wealth that is quite different from the way it appears in London, where even wealthy people's homes are smaller. Rather than wide lawns, you see lots of little paved streets. Mild, often cloudy and rainy weather allows a wide range of plants to thrive.

I soon learned that my co-founder in Beverly Hills also wanted me to change my personal grooming and 'look' to match the standard for this environment. My Instagram profile was the first thing I was asked to change. He advised me to match the visual landscape of LA and become more glam. This transformation went beyond clothes and accessories because in LA I encountered the beginnings of what is now known as 'the Instagram Face'.[9] The Instagram Face is a beauty standard shaped by social media. It has almost no blemishes, really big eyes, bud-shaped lips, a chiselled face (and obviously there's a very thin body to go with this face). I vividly remember a Getty photographer taking my picture at the gallery and then photoshopping my face to emphasise my jawline (that coveted 'chiselled' look!). I was twenty-three and remember thinking that's so weird because it's obviously not my face. The photographer had thought he was doing me a favour by putting that 'better' picture of me online, but it wasn't me. In many respects the person I was being asked to morph into physically was a representation of the visual aspirations that our world puts on women.

This phenomenon of being constantly shown an ideal face and body across advertising, social media and traditional media has since become an even bigger part of our lives beyond LA, in part because LA is one of the biggest sources of images that make up our shared visual landscape globally. LA, home to some of the world's biggest celebrities and media companies, is behind a significant proportion of the content that we consume daily, which now includes not just film and TV but also social media content, so its visual landscape influences all of us through our visual consumption. Most of the world watches content created or curated by Hollywood, so visual trends affect us all. For example, Netflix has 232.5 million subscribers around the world, of which only 30 million are based in the US and Canada and 17 million are based in the UK – leaving quite a large number in other countries! All of these viewers consume a collective 6 billion hours per month on the platform.[10]

Even when we are not in the public eye ourselves, we are constantly being exposed to what we should be looking like, where we should be living and how that should look along with the kind of life we should be aspiring to and living. This drives up our anxiety because you are never going to meet this constantly changing ideal, yet you are constantly targeted by visuals telling you, 'This is how you could be so much better.' What we see is the underlying influence taking over and people using that as a sort of guideline, even though they might otherwise have made different choices. I think Los Angeles and the Hollywood culture pioneered, sadly, part of this aesthetic we can see elsewhere. That said, I am grateful to have been exposed to this culture as it forced me to develop my visual critical thinking skills and see the reality behind the myth: how the people who were pretending the most were the most miserable.

Case Study
MTArt's first research project

To determine some common visual literacy levels, back in 2016 and 2017 my agency conducted two studies in London. I was on my own and didn't have a team yet so I stood outside London Bridge and Whitechapel tube stations where I had implemented two key public art projects by our artists Jennifer Abessira and Marine Hardeman with the help of our clients Network Rail, Team London Bridge and Tower Hamlets. I interviewed people for hours with questions that we had pre-determined with data analyst Vishal Khumar.

The idea was simple: how could we monitor the change in mental health, the creation of an emotional connection with the neighbourhood and an overall increase in happiness with the integration of visual arts in the everyday cityscapes? I used a mix of wellbeing metrics and the economical metric 'Willingness to Pay' that was used to measure how willing people who lived and worked in a specific area were to contribute to a new park or public art project. The latter was to enable me to understand the economic value that we were building for the neighbourhood. These hours interviewing people were long. I was rejected countless times but still drew in 800 respondents. We asked these people what kind of images they encounter on their commutes every day and found that they were mainly getting exposed to content that is materialistic, narcissistic and sexually based. They reported that this content had a negative effect on their wellbeing. When we crunched the numbers, we could see that for 84% of the people we surveyed, their sense of wellbeing increased substantially from starting their day by seeing our artists' work instead of a piece of advertising. And 82% of the people we surveyed responded positively to the idea of contributing financially to the creation of more public art projects in their cities. While £2 or £5 per person may not sound like enough to make an impact, if you add up the number of people who responded positively and extend it to a full neighbourhood, this can become a new way to finance more public art projects, or give politicians and local councils the courage and data to support these projects. This study demonstrated that we can feel the impact of changing our visual consumption, even on a small scale, and that this impact is significant enough that most of us would be willing to pay for it.

Exercise
What's Your (Visual) Story?

If you hadn't considered what your visual narrative is until reading this chapter, I do not blame you. It's something I wish I'd done early on in my educational journey. But, the more I learn, the more I think visual literacy should be taught on a par with writing and other types of literacy. To give you the opportunity to get to know your own narrative, I have devised a set of questions. Take a few moments to consider the following questions:

- What was the visual narrative you grew up with?

- What do your immediate surroundings look like on an average day?

- Do you identify yourself as a consumer of a particular type of media (online/books/magazines/ advertising/ TV series, etc.)?

- Do you gravitate towards particular colours or aesthetics?

- What visual narrative of yourself are you creating right now? How are you portraying yourself to the world and what are you telling yourself about who you are through the visuals around you?

- Is that the narrative you intended to create?

- Is this the narrative you want to identify with in five years' time?

Your visual narrative will always evolve. And the narrative we tell through those visuals will also change as we move through the stages of our life. You will make mistakes. We all do. That's fine. In your personal or professional life, certain visuals may be right for you for a while and then lose their appeal. The important take-away here is that we all create a narrative through the visuals we consume and the ones we share and create. The only variable is that we either do so deliberately or unwittingly.

Chapter 2.

We Live in a Very Visual World

Our Visual Consumption

It's morning. Your alarm is going off, meaning that unless you are a parent to a needy early-bird kind of human, you most likely have been awoken by your phone. And so, here it goes: it's morning, you reach for your phone, and since you're already holding it, you end up looking at . . . something. In a way, it doesn't matter what that thing is, your eyeballs are now glued to your screen; stimulating all sorts of hormones in your brain. Whether it's someone you hate-follow sharing a big piece of (good, or bad) personal news, a cryptic post by an ex of yours, or an email from your favourite brand, all that scrolling, all those images influence how you feel, what you think about and what you do. They have the power to create a new desire in you or trigger an emotion and act as visual clutter.

If someone was harassing you publicly, you'd probably be upset. If it escalated and they followed you up and down the street, you'd want to report the assault and expect some kind of action. Yet, to our brains, this never-ending onslaught is not totally dissimilar to being assaulted by visual messaging all day every day. With much of this content, we start the day stressed and anxious but don't realise that the images we see on our daily commute or morning scroll on the toilet are negatively impacting us.

The average person will be largely unaware of how this dynamic affects their nervous system and feeds their insecurities. Yet, meanwhile, the masterminds behind these visuals are very aware of their power.

The good news is that, in an age of digital curation, we can increasingly take control of them. It all starts with a visual audit. This and the next few chapters will offer ideas and insight on how to improve your visual experience:

- At home
- Online
- In our communities.

It can be hard to change our visual consumption habits because they can easily become unconscious and addictive. The visual audit allows us to step back and take a conscious look at what our consumption patterns are, to spot any harmful patterns. Our visual consumption and the harm it can cause can be compared to the early days of the tobacco industry. While for a long time tobacco consumption was simply considered to be a bad habit, we now recognise nicotine to be a highly addictive drug. The physical effect of nicotine and nicotine withdrawal on the body is significant, but the psychological factors of tobacco consumption are even more dangerous. Nicotine releases dopamine in the brain, causing mood-altering changes that make the person temporarily feel good. Inhaled smoke delivers nicotine to the brain within just twenty seconds. This immediate 'rush' is a major part of the addictive process.[1] We now know that consuming certain kinds of visuals, such as likes on social media, can give us a rush of dopamine too. Long-term changes in the brain caused by continued nicotine exposure result in nicotine dependence, and attempts to stop cause withdrawal symptoms that are relieved with renewed tobacco use. The physical withdrawal symptoms are the biggest reason why quitting tobacco is so hard, but what makes it even harder is that most smokers develop conditioned signals, or 'triggers', for tobacco use. For example, some people always smoke after a meal, or when they feel anxious. These triggers lead to behaviour patterns that can be difficult to change.

These behaviour patterns are similar to the addiction we have developed towards certain types of visuals, mostly visuals pushing us to consume more or to desire to be somebody else. Advertisers have long known how to use visuals to manipulate viewers. For example, Edward Barneys, a lesser-known relative (uncle) of Sigmund Freud, suggested to large tobacco companies that they should pay Hollywood celebrities to smoke in movies to inspire more people to take up smoking.[2] Bette Davis was reported to be paid $2,500 ($36,646 in today's money) as the star of the film *Now, Voyager*, while John Wayne took $1,000,000 ($14,000,000 in today's money) – don't get me started about the gender pay gap![3] These stars were paid to smoke on screen in order to make smoking seem desireable by extension. The same psychology can be applied to today's world with a constant flow of visuals sent to us all day long. A lot of the images we see are carefully crafted to spark desires inside of us; desires which serve the commercial interests of the image creator. It is important to remember that in today's attention economy one of these commercial interests is to get you to consume more visuals or remain on a platform where more advertisements can be shown to you.

The vehicles for distributing these images to us, such as social media feeds, are designed to be addictive. In the Netflix documentary *The Social Dilemma*, social media insiders and tech designers lifted the lid on the addictive properties of the visuals we consume on social media.[4] The film argues that social media is essentially an 'attention extraction model', where design features like infinite scroll are deliberately created to exploit our human desire for connection and validation. We get a little dopamine hit every time someone likes a post. No wonder so many of us are sharing visual material that we may not be entirely comfortable with. The need for the dopamine hit is too strong. But it also leads to elevated levels of anxiety when we present our show reel instead of our reality, which makes us feel bad, and then drives us back to social media as a digital pacifier of sorts.

Instead of connecting with other people in real life and pursuing our interests, we continue with the maladaptive coping strategy, an easy answer when we feel lonely, uncomfortable, sad or anxious. We feel compelled to scroll through visuals on our phones to get a quick rush of dopamine – the novelty of seeing something new or 'discovering' new information keeps us coming back to look at more and more visuals, but these visuals that feel good in the moment can have adverse effects. For example, you wake up in the morning and the first visual you see is a very slim woman, or a man who is really muscular, who you probably don't even know personally, doing exercise. A few days later you start looking at your body with a more critical eye and perhaps restrict yourself from wanting to eat what you want. Maybe you even download a weight-loss or fitness app.

Case Study
Florie-Anne Virgil

During the Covid pandemic lockdowns, journalist Florie-Anne Virgil was stuck in a very commercial London neighbourhood in zone three, surrounded by billboards with lots of adverts by retail clothiers with big flashy images on them. She's someone that loves culture, loves going to the theatre, loves going to the movies, loves going out into nature. Usually, this means that her visual consumption is varied and that any stressful and overly commercial visuals she encounters are balanced out by more peaceful and creative visuals. During the Covid pandemic, all of this was restricted. In the absence of her usual varied visual consumption, she found herself addicted to the local and world news television programmes. Day in and day out, she watched the news, alternating between visuals about the world's most pressing and difficult current affairs problems and the advertisements in the commercial breaks that were pressuring her to want things she didn't have. When she went outside, instead of the reprieve of nature, the commercial landscape around her felt like an extension of the commercials between news segments. She simply felt bombarded by commercial visuals when walking to or from her city house with absolutely no escape, only to be bombarded with these visuals at home too. There was no cultural event or anything that she could escape to, apart from those adverts. Being home constantly meant she was watching the television a lot, but it impacted her negatively with shockingly graphic visual news such as reports of the Syrian war and ensuing diaspora to Europe.

It was negatively impacting her wellbeing so much that she decided to move elsewhere with her boyfriend. They made a total change and got a tiny place in the countryside and got away from the city environment entirely. When she told me about her experience, we were sitting in her small garden in Kent, surrounded by flowers and trees, feeling the breeze on our faces, and she was wearing a long yellow tunic. She looked so relaxed, as if her new visual environment had removed any anxiety from her. As she spoke about her need for change, I couldn't help but look around me, observe the details of the leaves in the trees, look up to the clouds and with her welcome this change.

The negative impact on her emotional state that Florie-Anne noticed when her visual consumption changed is proven to happen all the time. In 2018 the University of Pennsylvania conducted a study with 143 undergraduates who were randomly assigned to two groups, one with limited access to hyper-edited and perfectionist social media content and the other asked to continue their usual social media with hyper-edited content use for three weeks. The limited group showed significant reductions in loneliness and depression during those three weeks over the group that continued using social media.[5] An earlier study at Columbia University in 2016 found that a brain network responsible for suppressing behaviours like inappropriate or unwarranted aggression became less effective after study subjects watched several short clips from popular movies depicting acts of violence, which could lead people to lose their ability to control aggression.[6] These two studies combined show the pressing urgency of being aware of the visuals we consume daily and their quantity because they can have a serious affect on how we feel and how we act. Thankfully, there is a way out.

Exercise
The Visual Audit

So, what can you do? The first thing I'd encourage someone looking to make a change is to conduct a visual audit. Instead of keeping a food journal for a week to track what you really eat or a money journal to track what you really spend, how about keeping a visual journal to track what you really see for a week? This exercise will help you identify the kinds of visuals you are consuming. From there, you will be able to better identify how those visuals are affecting you and make conscious changes during the Visual Detox. Take your notebook everywhere – or whip out your 'Notes' app – and record what you see. If possible, take photographs of the things you notice too. Take a moment to consider how you feel in each moment as you record your experiences. Was there a change in how you felt before and after your interaction with the visual? Did what you see change how you felt? In learning your triggers, you can learn how to steel yourself against such a big emotional response.

YOUR VISUAL AUDIT

1. What is the first thing you see when you wake up?

2. Where in your home do you spend the most time? What does that room look like?

3. Do you commute to work the same way every day? If so, what do you see during your journey?

4. Do you spend any time in nature? If so, what do you like to see?

5. Do you have a favourite spot? A café or bench in a park? Take a photograph of that place. Why do you love it visually?

6. Do you walk with your children or take them to school? Do you take different routes? Why do you prefer a particular route over another?

7. If you often use public transport, make a note of what you see, uplifting, dull or anxiety inducing. Record it all.

8. What do you watch on TV? Do you stream? Are you a gamer? Is there a favourite TV show that feels addictive and you watch it anxiously?

9. What do you see on social media? What type of feeds do you follow? Do you notice the difference with a particular type of content on your mood?

10. What is the last thing you see before you go to sleep?

If you're likely to forget to record your visual consumption, try setting an alarm on your phone – perhaps one of the few times when I'll recommend you get on your phone in this book – to sound every few hours. Make a note of your visual intake each time the alarm sounds and record how you feel. The idea is to become more familiar with what you are seeing and the impact those visuals have on your wellbeing. Once you've conducted your audit in your notebook and have taken photos of those visuals, look through them one by one and consider what you've learnt so far.

I did the exercise and while I knew the issue, I had missed so many visuals that I was consuming and was not aware of. I realised that when I mindlessly scroll through my social media feed, I tend to consume a lot of content about stay-at-home mums and their lifestyle. I find myself targeted by their content since I became a mother and there are so few representations of female leaders raising their kids online. It's always an uneasy feeling when faced with what society wishes you to look like visually versus what you are. When I feel like this part of my visual consumption is starting to affect me negatively, I make a conscious effort to bring my focus back to visuals that reflect and echo a visual landscape that is more closely aligned to my own life so that I am reminded of my goals, achievements and priorities instead of distracted by what I am not.

Exercise
Remembering What You See

An easy way to remember the various visual elements that converge to influence us is the technique used by journalists to uncover more of the key details of any story. In the same way we can use the same technique to simplify the key elements we should consider regarding our visual consumption. The technique is known as 5W1H and it refers to:

- What (visual)

- Who (human)

- When (human + environmental)

- Where (environmental)

- Why (visual)

- How (environmental + digital)

With each image you've collected from your week's observation ask yourself:

- What am I seeing here? Is this visual deliberately conveying a message to me that I may not be fully aware of? Think about the size, colours used, composition, orientation, the use of props, source, intention and veracity. Usually a colour, a display or a composition would have attracted your eye.

- Who am I and how might my upbringing, culture, ethnicity, socio-economic background and education potentially interact with this visual to impact the meaning I draw from the visual? Could my past conditioning and natural bias be influencing what I see? Or is it influencing the meaning I give to what I see? Who created this image, and what was their intention; what do they have to gain from it?

- When did I see the visual? Was I in a hurry to get somewhere? Also, what stage of life am I at? Does that influence what I see and notice?

- Where is the visual? Is it indoors or outdoors? Is it a billboard or a building? What is the 'where' telling me about the visual?

- Why am I seeing this? What is the person who paid for this or created it trying to get me to do? Do I really want to do that?

- And finally, how am I seeing this? In real life with my own eyes, or am I seeing pictures or videos on social media or TV?[7]

Some of these questions might not immediately have clear answers, but the next few chapters will help give you the tools to answer them for any visual that you encounter in your life.

Chapter 3.

Understanding Your Visual World

When speaking to friends who work across other fields, I am often asked, 'Okay, so, what should I be looking out for in a visual cue?' This is when I usually get to be my nerdiest, unpacking the visual signifiers underpinning some of the most popular artworks and adverts in the world.

So far we have been exploring our visual narratives, from birth to adulthood, and how they impact us. Now it is time to unpack how the visuals that have shaped us work and the impact of specific kinds of visual messages on our health and wellbeing. We will look at the secret or hidden codes of visuals. This includes the messages we receive from what we see (or the visual information itself), how who we are interacts with that information and how the environment also plays a part. Where these three spheres – messages received, the visual itself, our interaction with it – intersect represents the unique meaning we draw from our visual consumption.

How Do We See?

Visual narratives are about 'what we see' with our eyes. But this also begs the question, '*How* do we see?' Apart from when we are sleeping, for those of us who can, we start perceiving the visual world around us from our birth until the moment we take our last breath. Once the light comes through the pupil it will go through the eye's lens. Just like in a camera, the lens is used to focus on an object and direct the light to the back of the eye. Tiny nerve cells are able to take the electric form of the image in front of you and send it to the brain's visual cortex or 'vision centre'. The vision centre is located in the back part of your brain (the occipital cortex or lobe) where it decodes the electrical information coming from the retina. Neuroscientists have discovered that most of what goes on in our brain is preconscious, which means that it comes into our brains without our conscious mind realising it.

When the brain receives all of this information it then needs to decide extremely quickly what parts of all the visual information it receives are important enough to be noticed by our conscious brains. Visual signals like shape/colour/contrasts in general are detected first, but they are sorted with visual attention bias, meaning that the brain focuses on things that are connected with memories or experiences which trigger feelings, and relate to things that matter to you right now, and discards other visual stimuli unless you consciously focus on them. This effect is called saliency and it is a form of natural filter that protects us from sensual overload. It can however lead to biases in perception and information processing that we are not even aware of.

This natural filter is necessary because we receive more information from the world around us than our brains can process. Scientists have actually calculated the bandwidth of human consciousness: researchers from the University of Pennsylvania discovered that the retina transmits about 11 million bits of information to the brain per second. That is about the capacity of an ethernet cable. A German physiologist named Manfred Zimmermann calculated that our other senses transmit another one million bits of information to our brain.[1] It is worth pausing for a moment to recognise the difference in this context. Our maximum capacity is 11 million bits per second, which is pretty impressive. Unfortunately, of that massive flow of information, no more than 40 bits per second actually reach consciousness.[2] Bits are the units of information that computers use to process, but to put it into perspective, that is like our senses perceiving an entire fully grown oak tree with all of its leaves, but our conscious mind only being able to take in one single leaf. Considering the ratio of input from the five senses it is therefore very likely that our visual consumption is having a significantly bigger impact on what reaches consciousness than any other sense.

This scientific phenomenon is a crucial part of our life and experience, so much so that we stop paying attention to the fact that not everything we see is registered consciously. The late novelist David Foster Wallace, in his commencement speech to the graduating class of Kenyon College in 2005, said that a fish doesn't realise how important water is for its day-to-day experience as it swims around its pond; in other words, it can be hard to realise how the things that surround us every day are affecting us.[3] We barely realise how important our visual consumption is in our day-to-day experience because we aren't consciously aware of most of what we see.

Case Study
Blind spots

The real-world impact of just how little information gets into our consciousness was brilliantly demonstrated by Daniel Simons of the University of Illinois and Christopher Chabris of Harvard in 2022. Their study was simple: they asked a group of 3,000 volunteers from eighteen to sixty-five years old to watch a recording of a basketball game and count the number of passes one of the teams made.[4] They were also asked to report anything else they noticed. A few minutes into the recording, someone dressed in a gorilla suit walked onto the court and wandered through the players for several seconds before turning to the camera, beating its chest and walking off camera. And yet half the participants didn't see it. They had narrowed their conscious focus so much to count the number of passes the players made that they failed to see something as obvious and unusual as a person in a gorilla suit. Many refused to believe this had happened and demanded to see the recording again.

What's more, we might assume that when 100 people see an image or visual, all 100 people see the same thing, but that isn't the case. The way we receive messages from visuals is different for each of us. Sometimes a message is affected by our subjectivity, sometimes it is intentionally impacted by the composition or characteristics of the visuals themselves and sometimes the message is altered by the environment it's viewed in (more on this in chapter 4). As Claire Harrison explains in 'Visual Social Semiotics', the meaning that those 100 people extract from the visual is a 'negotiation between producer and viewer, reflecting their individual social/cultural/political beliefs, values and attitudes'.[5] In other words, who we are shapes what we register in our visual landscapes as well as how we interpret the elements that we notice.

To understand what you see, you have to understand what your circumstances are bringing to the visual. The visuals we see daily are driven by human dimensions and variations that can and will influence interpretation. In the work I do as a visual expert, I have identified the following characteristics:

- Conditioning
- Attention
- Timing
- Emotion
- Education and experience

As a mnemonic tool, you may wish to remember these elements as 'CAT-EE'. Let me run you through each of these characteristics:

Conditioning: The Impact of Nurture

In his remarkable book *Sapiens*, Yuval Noah Harari states, 'Humans emerge from the womb like molten glass from a furnace. They can be spun, stretched and shaped with a surprising degree of freedom. This is why today we can educate our children to become Christian or Buddhist, capitalist or socialist, warlike or peace-loving.'[6] A huge part of the spinning, stretching and shaping process is determined by conditioning. As Harari alludes, human beings are not – so to speak – 'complete' when they are born. Right from the start a baby is 'downloading' vast amounts of information through their five senses, even in instances where they may have reduced faculties. That information is then used, albeit unconsciously, to work out what is right, wrong, good, bad, acceptable and unacceptable as they find their place in the world. They also use it to identify potential danger.

This download is an innate learning process that allows us to absorb vast amounts of information, knowledge and data that we then use to develop a map of who we are, what we are capable of, how to get along with others and get ahead in the world. The problem is that most of the processing is done when we don't yet have a fully functional neocortex and, therefore, we don't have the cognitive capability to determine accuracy or validity of the information we use to spin, stretch and shape our lives into the person we become. That part of our brain isn't finished until we are in our early twenties. This means that as adults, we sometimes have to do the difficult work of unlearning ideas and beliefs that feel deeply ingrained in who we are when we encounter new information later in life.

Case Study:
Life in Hong Kong with Theo Ruppert

Hong Kong is one of the most densely populated places in the world, which has huge housing issues. With 17,311 people per square mile, or 6,659 people per square kilometre, it's the second most densely populated metropolis in the world.[7] Life there is densely made of commercial visuals, retail visuals and architectural visuals. Photographer Michael Wolf creates highly detailed large-scale views of high-rise structures that are homes to millions in modern Hong Kong. His 'Architecture of Density' lens captures the concept of urban density in astonishing (and frightening) detail. Looking at his work can help give you an impression of what it feels like to be in that visual environment.

Your sense of space, detail and scale is very different when your own visual space is so densely populated by people, adverts and thousands of visual distractions like city lights, billboards, etc. Financier Theo Ruppert lives in Hong Kong and recalls being bombarded with visuals when growing up. As we visited one of the small islands next to Hong Kong together, he described how, as an adult, he is used to almost blocking out a lot of the visuals which speak to him daily on his way to work. He cannot escape the hundreds of visual adverts he sees every day so he learnt to block them mentally instead.

I remember thinking how different this was to my

upbringing as I contemplated the sea for hours as it was my only visual distraction. As a child, my visual narrative felt like a blank canvas. While I had to imagine a visual narrative from scratch, Theo had to go through the reverse process and edit and reject the ones he didn't want to become. Those differences will mould human beings through visual conditioning and so, as we learn about our own visual conditioning, it's important to deepen our empathy with other visual environments to avoid jumping into judgements and conclusions too quickly. That's why when we see an image or scroll through a social media feed, we are viewing that visual consumption from our own unique perspective. There is no other you. No one grew up in exactly the same way you did, not even your siblings. Your experiences are yours alone, what you made those experiences mean is a product of your conditioning. That introspection can only happen if we first appreciate the way we are built. With the benefit of a developmentally finished neocortex (assuming you are over twenty-one) you can then use that part of your brain, especially your frontal cortex, to re-examine some of the assumptions you automatically rush to as a consequence of outdated and inaccurate conditioning.

It's impossible to know for sure what my life would have been like if I had grown up in a different kind of environment, but I think that if I had been bombarded by visual adverts like Theo was when I was growing up, perhaps I would have been unable to see and project my unique vision for my sector. Even if you thrive in a visually dense environment, I believe that creating a space in your life for 'blank canvas' moments helps you think more clearly and be more creative. If you find yourself constantly surrounded by new pictures and images all the time, try finding a few minutes to sit in a visually empty environment without any distractions to give your mind a break from having to process all of that information.

What's On Your Mind?

Your brain is able to process images thanks to the four lobes of the neocortex, different regions with specific functions that work together – the frontal, parietal, occipital and temporal lobe – and makes up approximately half the volume of the human brain. It is responsible for the neuronal computations of attention, thought, perception and episodic memory.

The frontal lobe is responsible for the selection and coordination of goal-directed behaviour such as task switching, reinforcement learning and decision-making.

The parietal lobe is believed to play a role in decision-making, numerical cognition, processing of sensory information and spatial awareness.[8]

The occipital lobe is responsible for visual function and hosts the primary area for visual perception which is closely surrounded by the visual association area.[9]

The temporal lobe houses the hippocampus and the

amygdala. Among its functions are processing sensory information and deriving language, emotions and meaningful memories.[10]

The neocortex itself is a structure with great information-processing and -storing capacity. In humans and possibly other mammals, the neocortex mediates consciousness. Most importantly, the neural circuits in the neocortex are modifiable in development and throughout life, so that species become specialised for relevant perceptual and behavioural abilities, and individuals acquire individual skills, abilities, personalities and memories.[11]

These neural circuits create reflex responses and natural bias. Can you remember the last time an image made your blood boil and before you could stop yourself you'd flown into a rage? Within minutes, you may have become bewildered as to what happened – that is a typical case of reflex response. Something about the interaction fired up your automatic fear response and you reacted before your thinking brain could fully engage.

The Role of Bias

As for natural bias, almost all humans prefer to work and interact with people who look like them, share their values and who, it seems, are like them in some way. This is why there are so many middle-aged white men in the upper echelons of business – natural bias means they are instinctively inclined to be drawn to and to value (and hire) people who are like them over those that are not. Most of the financial pages of *Forbes* magazine, up until a few years ago, featured suited-up white men, but with gender biases slowly being broken down over the past few years we're starting to see a change in who goes into business with an increase of 21% in female entrepreneurs.[12] However, while there's been an increase in the number of female entrepreneurs across sectors, most of the financial pages still feature dispro-portionately more suited white men than other kinds of entrepreneurs.

Another example of bias is that teaching staff in the UK in 2022 were 75.5% female, in nurseries and primary schools women even made up 85.9% of the staff.[13] This is caused by the fact that biases in society make people think that women are better suited to caregiving roles, meaning that they are encouraged to take those roles and more readily hired over men.

One of the reasons societal changes around gender roles are happening slowly is that many of us were born into a world where the primary visual narrative in our cul-ture defined gendered roles for men and women. We grew up watching female characters play the part of caring role models on our favourite cartoons and on advertisements, reinforcing the idea that caring roles are feminine, and we watched male characters carry briefcases and fight bad guys, leading us to repeat those patterns.

Visual bias is still playing a strong part in who does what in our societies. My second son Vivaldi was born in March 2023, and when he was three weeks old I gave a talk about investment in the art world, standing with him on my chest in his sling. It was one of my most viewed videos for months, with many women commenting that they had never seen anything like it. We are not used to visuals that mix messages about business and leadership with parenting and caregiving, but many people are engaged in both every day.

Bias is about far more than gender roles. Unfortunately, class still plays a role in this discussion right alongside other socio-economic mobility or gender access issues. For instance, I'm in a very nice area of London now which is specifically designed as a very picturesque type of neighbourhood. In practice, this means there are few to no chain stores, on purpose. They just bring in the boutique-y letter brands, but this has also put the rents higher. If I was in a low income area (and I have been), I'd see the total opposite, with the big supermarket signs flashing everywhere. It happened for me, so I do figure that a virtual environment geared to one's class or background affects you. People make judgements based on what they perceive to be visual markers of class, which affects how they move through the world.

By taking control of our visual consumption, we can learn to overcome the biases we absorbed growing up. We can change our visual narratives by seeking out visuals that challenge our preconceived notions and widen our horizons. There is a slow shift taking place in our mainstream visuals around diversity, opportunity and the realities of 'real people's' lives. I think we are in a pivotal moment of seeing a diversity of role models. For example, the Nike adverts with Serena Williams getting back out on the court and hitting the ball after she had just given birth, and showing hijabi athletes, are making different images of what an athlete looks like part of our everyday lives, expanding our definition of who can be an athlete.[14] It goes beyond advertisements – real world examples are important too. Seeing a woman in the Nike advert running the Boston marathon, and being the first female to do so, is visually really impactful for changing our definition of 'women': women can perform, look muscular and challenge the gentle image that is visually bombarding all of us every day. Everytime I see Serena hitting that ball with so much strength on these ads, I am emotional because I know that it will trigger so much change in who we can become as women and who gets to represent us.

In short, the environments and expectations of our upbringing will impact what we pay attention to and also how we interpret what we pay attention to, and messaging that prescribes certain roles for different kinds of people can limit the opportunities that are available to them as well as the ones they seek out. Even for people growing up in the same country, someone from a middle-class home will be shaped by a different visual landscape to someone growing up in a council estate. I don't want to make this book about class, Mike Savage already does that in his book *Social Class in the 21st Century* but it's important to fight against a monopoly of visuals as we all grow up in very different visual environments.[15] In my sector, where 90% of people come from privileged backgrounds,[16] this means that the visual biases in the art world will be skewed by shared biases. I wanted to challenge these biases by bringing in new points of view through my agency. Most visuals we consume are still too similar and aren't speaking to all of us. French journalist Judith Duportail's book, *Love Under the Algorithm*, revealed findings that 72% of the accounts that performed best on Instagram represented a wealthy Western world lifestyle.[17] This underscores that many of us are excluded from the dominant visual narrative or do not see ourselves in the main characters of those stories because the majority of the imagery we're seeing is not about us.

Case Study
Àsìkò

Visual artist and conceptual photographer Àsìkò has spent a lifetime crafting a visual environment which blends Nigerian and European culture together. Born in London, UK, shortly before moving to Nigeria aged two, the artist describes his earliest memories of the country as being bestowed with a honeyed hue – 'everything, from the sky down to the floor was lit in a warm tone' – as a result of the combination of dust and red clay. This 'warm' energy transcended into the architecture, which despite carrying Western aesthetics, he recalls as being distinctly Nigerian – due to its embracing of the country's heat, with its large, open space and lack of architectural ornamentation.

Inside the home, the artist was fascinated by the vast number of objects that filled the interiors, acquired by his father – an avid collector of Nigerian art. Shelves, cupboards, tables and floors were lined with a diverse array of art, including sculptures, tapestries and paintings, depicting scenes of African city and village life. He recalls his father coming home weekly with different African antique furniture; there was always a new addition to the environment. One specific room in his house, which mimicked an Italian *studiolo*, contained all of his father's greatest curiosities. It was a place that carried a kind of 'sacred energy', where objects were 'treasure' and, as such, had to be treated with the 'utmost respect', thus when entering he was mindful not to touch, but to treat them with a sort of divine spirit. It was from these early observations that Àsìkò began to witness the value of art. Yet, it was also in unexpected places in the home that he developed an affinity for different art forms. He recalls that objects in the home that were used for functional purposes, such as wooden serving spoons, bowls and plates, amazed him for their 'small, intricate details'. It was an accumulation of all these things which informed the artist's early visual diet.

Upon returning to the UK aged sixteen, he became starkly aware of the visual contrasts between Nigeria and London. He describes the latter as carrying a 'colder tone' with everything feeling smaller and more concentrated. From his deep interest in the art collected by his father in his youth, the artist was conscious of the different visuals surrounding him in London, and more broadly in Europe, and was particularly taken with the Renaissance art of sculptors such as Donatello (1386–1466), and artists like Leonardo (1452–1519) and Michelangelo (1475–1564). Following a seminal trip to the Vatican City, where he witnessed the Sistine Chapel, Àsìkò was enamoured by the epic scale, beauty and drama of Michelangelo's friezes. It was the robust, dynamic forms present in *The Creation of Adam* (*c.* 1508–1512), which had a particularly profound effect on him – '[the work] possesses a spiritual connotation', which although is 'different from African art, carries similar themes'.

Yet, within this visual environment, with its depiction of beautiful, powerful figures, he felt there was something missing. Aspects from Nigerian culture, such as the embrace of the Black nude female form, were absent. Thus, when creating his own works, he decided to subvert this. In his *Madonna* (2020), part of his 'The Adorned' series, the artist decided to frame the black Madonna and child like a Perugino. He takes the gilded haloes found in the works of Cimabue and Giotto and mixes them with the textures and patterns found in the arts of his own Nigerian culture. It is something rarely seen in Western art, especially not in the art of the Renaissance – the hybridisation of European and Nigerian culture. The subsequent result crafts a new visual narrative – one which puts the beauty of the nude Black female woman to the fore.

When creating his public artwork *Of Myth and Legend* (2022) for the The Knightsbridge Estate, composed of multiple billboards on the bustling streets of Knightsbridge, London, he wanted to shine a light on African heritage and give greater representation and autonomy to the Black community. The result of the project irrevocably changed the visual environment of the area. Thousands of people witnessed, perhaps for the first time in a London setting, the Black body being portrayed as a mighty deity. As he notes, 'It's important in London, a space with so many different ethnicities, to see yourself in it.' It is this blend of putting African culture into a European context which has resulted in new visual narratives being created. Àsìkò has crafted a visual environment in London which is both inclusive and representative of the population as a whole.

Exercise
Spot the Bias

Too often we consume visual information without ever stopping to consider how our biases are influencing the images we seek out and how we interpret those images. The next time you look at something or an image grabs your attention on TV, or advertising, consider what is drawing you in:

- What conclusion or assumption have you jumped to?

- Could natural bias be playing a part?

- Is the use of colours or composition different to what you are used to?

- How does the image make you feel? Is there a strong emotion? What is it?

- Also consider which perspectives are included in your visual consumption. Are the images you see created by people like you? Do you see people like you represented in the visuals you consume? Are there any voices missing from your visual consumption?

Attention: The Importance of Focus

We have more access to more visuals than any other generation before us. This has meant that we are experiencing new challenges when it comes to curating what we pay attention to and how we stay focused on the things that matter most to us. In her work studying how constant access to the internet has affected our attention span, psychologist Elena Medvedskaya found that the frequency and duration of internet use were negatively correlated with attention span, meaning that people who spent more time on the internet tended to have a lower attention span.[18] The study also found that specific internet activities, such as social media use, were associated with lower attention span. Research like this indicates that being mindful of our digital consumption in particular is really important for protecting our mental wellbeing. Visual distractions are everywhere: how many times do we find ourselves watching the television with our own phone in our hands now? We end up having visual distractions from the walls of our own home surrounding us to the TV broadcasting our favourite show and all the while being on social media apps. As we get more accustomed to intense visual stimulus, we can find outselves seeking out more of that intensity.

That isn't to say that there aren't healthy ways of interacting with media. I keep a close eye on my eldest son and his consumption of television and other visuals. I feel that the adverts and visual culture aimed at kids are not only getting brighter in terms of colours, but also faster in terms of speed of delivery (talk fast, talk briefly, move on) and action (each scene seems to be high-speed action). These intensified visuals are also becoming very common in media consumed by adults. Viewing these kinds of intense visuals has an effect of constantly triggering different emotions. If you were to compare the very early Disney animated films, first of all, the colours are much lighter, paler. Secondly, the speed of the action and the visuals (due, I agree, to the technology at the time) were at a much slower pace or speed than the action that he'll be looking at in more modern animated film. As another example, when you scroll through social media, you'll find yourself quickly moving between visuals that trigger a wide range of emotions, from the sadness of natural disasters to clips of comedians performing their funniest jokes. For both children and adults, it's no surprise that the increasing intensity of emotions is also increasing our emotional and physical response. But don't misinterpret my comments: I really don't love the idea of flatly stating, 'This is something good, this is something bad.' I will always go for building awareness. Whatever your reaction to or awareness or reception of old-school or modern cartoons, you're not going to ban and cancel cartoons. You're going to ask yourself what the effect will be on you or your children and ask, 'Do we want this for them or us?'

Once you've got this awareness, just balance it out for yourself. If my son does one hour of less-stimulating cartoons, we just go to the park. Or I get him (and myself, actually) contemplating something in nature for a period and focus on it as a sort of meditation to help move the sensations down a notch. By moving over to something that will feel very different and contrasting to their experience, I've created some balance for him (and myself). It's all about balance rather than me constantly saying, 'This is good, this is bad.' I don't like the idea of telling people what to do. I would rather you develop awareness allowing you to decide what you want to do, and after that you're aware of the whole of it. Build your awareness.

Advertisers, politicians and influencers know the power of images and visual narratives and they use them because they are more likely to get noticed. But is what we are noticing helping us, stimulating us, challenging us or hindering us? The next time something captures your attention, allow yourself to be consumed by the visual or story but then ask yourself what it was that caught your attention. Are you aware of how the visual made you feel? Did it make you feel a specific emotion – fear or anger perhaps? Is that helpful to you?

Timing: A Crucial Factor

There are two parts to the timing component:

First, when we consume the visuals we are exposed to will also impact how we react and interact with these visuals. If we are rushing to work, moving down escalators to get on a train, we are focused on getting to the platform on time. Once on the platform we may be more relaxed and therefore more aware of what we see. It won't surprise you to hear that this is why the ads are larger on the wall behind the train – or whatever form of transport – than they are on the walls as you travel down the escalators to the platform. Wherever people congregate but are waiting for something else to happen offers the most eyeballs and potentially the most bang for the advertisers' buck.

The opposite is also true: sometimes we encounter a visual in a time or place that makes it harder to absorb the message. Sometimes a visual comes to us when we're preoccupied with thoughts or emotions that alter how we receive the visual message. Put simply, have you ever overreacted to a message as someone happened to text you negative news at a time when you felt particularly tired, down or stressed? Well, visuals are exactly the same, they can catch you at a moment of poor timing too and, as a result, you may overreact and act in a way that you would regret later. For instance, buying a piece of clothing that you do not need, feeling insecure or provoked by an image, etc.

Secondly, consider the impact of timing with regards to

the stage of life we are at. Since our brain constantly has to filter and prioritise what we pay attention to, we will be more alert to visuals that relate to our current life experiences. Humans always have an innate awareness of danger but beyond that, what cuts through our consciousness is influenced by what is most relevant to us at any given time. In 2019 I became a mother for the first time. It seems to me now as I move around London that the entire advertising industry is now dedicated to talking to mums. Is it that suddenly messages around motherhood have skyrocketed or is it that as a new mum, I am just aware of those messages for the first time? In all likelihood, I was always being bombarded with this information but only started to notice or comprehend it after having a child. Life is impacting what forty bits of information my consciousness is fully taking in. As I'll explain, our age, stage of life and what's going on in our life are also influencing how we interact with our visual consumption to create our visual narrative.

To get an idea of how cities and their architecture influence us at different ages, think of the London Underground. It was built with the average man's height of six feet as the builders' reference. If you're thinking of that reference now, when you go down into the underground, and consider that the design of everything is in relation to that measurement, you can put yourself into a new visual perspective, whether you're six-feet-six inches tall or a three-foot child. Now consider that the whole city has been largely designed around the average six-foot man. First of all, imagine that you're thirteen years old and are not yet at this full height (and may never be), walking around the underground, seeing enormous adverts – these might appear much more intimidating and loud to you than if you are a full-adult size. So I think breaking it down to the perspective of children, teenagers, is simpler. They don't have the standard size in that sense.

I also feel that our awareness of our 'normal' changes when something happens to us. As an example of what I mean, I broke my leg last year and had a plaster cast for four months, so I had crutches. I was not as readily mobile or as comfortable in my body or with people's movements close to me. I also found myself becoming much more aware of other people's disabilities or limited ability for movement, and of objects and people around me that might bump or hurt my leg as it healed. My awareness was suddenly heightened for my body's (absence of) safety. It's the same with biases, the same with emotions and feelings. We see what we are aware of. That's because it's just like anything, if you are conscious about something, you can see it more and feel it more acutely.

Exercise
Travel Through Time

Think about how your interests and needs have changed throughout your life. Are there any things that you didn't really notice as much when you were younger but are standing out to you regularly now? Next time you go on your usual commute to work, imagine you are seeing everything as a younger version of yourself – what stands out to you based on what you needed and thought about most at the time? Repeat this exercise imagining yourself at different stages of your life and see what changes – what do you see when you imagine yourself as a new parent? As a new homeowner? As a retiree?

Emotion: The Hidden Factor

How we see and interact with an image or engage with a particular visual narrative will also depend on another human element – emotion, and how we feel when we consume that image.

A must-read for students in France is the famous book *In Search of Lost Time* by Marcel Proust, in which he describes in great detail how his childhood experience of eating madeleine cookies was full of wonderful multi-sensory memories.[19] The cookie could 'erase' all the disasters of life in just one bite and take him back to easier, happier times. The entire passage (many pages!) is written in such a way that you can see it all with him. It's beautiful. We call it 'the madeleine of Proust' whenever we experience a sensory experience that brings us back to the past. This shows how tightly connected our emotions are to our senses. In my mid-twenties, I was unable to walk through or see images of the Borough Market area in London as it reminded of my break-up with someone I used to love deeply.

I think we should talk here about 'poverty porn' which is the opposite of Proust's delightful experience with the cake; even pronouncing the words is horrible and uncomfortable.[20] We can feel deeply from witnessing both the pleasing and the repulsive, and visuals can convey so many more emotions than words alone can.

One of the ways image-makers target our emotions is through 'poverty porn' – images of destitution that are staged to emphasise suffering to draw you in and make you more receptive to their message. Do you remember the horrible images of poor kids clearly starving on fund-raising adverts on television? They show the thinnest young kids with horribly distended tummies from both malnutrition and starvation, with runny noses, all surrounded by dust and dirt? I don't encourage you to look at anything like this, because just thinking about poverty porn makes you feel sick. But using poverty porn in those fund-raising adverts shows you how far an emotion can be triggered visually – and get you reacting the way the advertiser wants – namely by sending them a donation.

'Little Girl and the Vulture' by Kevin Carter, first published in *The New York Times* in 1993 under the headline 'Sudan is Described as Trying to Placate the West', is a Pulitzer-prize-winning timeless visual expression of poverty as it still exists in our world. Do you find it shocking? Does it make you uncomfortable? That was the goal (at least we can presume so) of the photographer and of the publisher. Its presence on the broadsheet drew readers to the accompanying text by showing them the high stakes of the issue.

As another example of how emotion shapes which images we're drawn to, consider the now-renowned June 1985 *National Geographic* cover by photographer Steve McCurry, 'The Afghan Girl'. The blue-eyed, dusty girl stares out at us as if to say, 'I dare you.' But what is she daring us to do, especially when we put her question in the context of the years- and decades-long warfare in her country? Then you read that Sharbat Gula was, at the time of the photo, an Afghan refugee in Pakistan during the Soviet–Afghan War. Now that you know she was a refugee, what does the image do to you emotionally?

It's usually what's most triggering that goes viral. What tends to become viral is what's most emotional. What triggers us most emotionally, activating our full body response, tends to stick in our memories and be something we want to share with others. Keep in mind that not all triggers stir up negativity or discomfort. While the visual could be triggering something that's positive for you, it nonetheless needs to be emotionally grabbing for it to first, get our attention, and secondly, emotionally trigger us.

There is a two-way feedback loop between images and emotions. Not only can an image or visual change the way we feel, by either triggering something that was created in the conditioning process or the image simply making us feel something by itself, but how we interpret an image is often influenced by the emotion we are experiencing when we view the image. If you are unhappy when you scroll through your social media feed then images that, on a different day, may have made you smile, will not. In fact, everything you see will be viewed through your emotional state at that time. And whatever state you arrive in is only likely to be amplified, not changed.

The scientific explanation for why images are linked to emotion lies in your limbic system. It is a collection of structures involved in processing emotion and memory, including the hippocampus, the amygdala and the hypothalamus. The amygdala is the brain's centre for emotions, emotional behaviour and motivation. Emotion is much faster than cognitive decision making. Up until World War Two, philosophers like Jean Paul Sartre used to believe that the thinking brain was in charge. According to Sartre, information would come into us from the five senses and straight to the thinking brain or neocortex.

Unfortunately, it's not that simple. Joseph LeDoux, a neuroscientist from the Centre for Neural Science at New York University, discovered that we have a neurological panic button.[21] When we receive the signals and information from our five senses (heavily weighted towards vision) a portion of the original transmission goes to the amygdala *at the same time* as the rest of the message goes to the neocortex. The main function of the amygdala is in emotional responses, including feelings of happiness, fear, anger and anxiety. This area is also key for the formation of new memories.[22] The amygdala interacts with the hippocampus by attaching emotional content to memories. Stimulation of the amygdala causes intense emotion, such as aggression or fear. This is why people jump into rivers to save children before they 'know' the child is in danger. The amygdala has triggered action in the face of a threat the thinking brain isn't even aware of yet!

As we saw earlier, conditioning is a learning process that allows us to classify experiences and alert us to potential danger. And emotion is the language of the amygdala – it is responsible for how we feel our feelings, especially fear and anger, controls our aggression and helps us to store memories. Whatever we may subjectively perceive as threatening or scary will result in amygdala activity. Many neuroimaging studies have demonstrated that patients with PTSD have greater amygdala activation and even the presentation of trauma-relevant words increased amygdale activation in PTSD cohorts.[23] Often our response to a visual is not a rational response, it's an emotional amygdala-based response to long-forgotten conditioning that flagged danger when perhaps none really existed. Not only do we need to recognise that what we feel is going to impact how we interact and engage with what we see but what we see is also, potentially, triggering negative feelings and fear responses that are not about the present moment.

This is why most of what becomes viral is negative – it's activating the fear response. This phenomenon has a name: the culture of outrage. In 'Amplification of Emotion on Social Media', Goldenberg and Willer argue that the culture of outrage is fuelled by social media algorithms that prioritise engagement and sensational content, leading to a cycle of outrage and amplification.[24]

Outrage has become a prominent feature of online

communication and it has negative consequences. Exposure to outrage on social media can lead to emotional exhaustion, decreased job satisfaction and reduced work performance, as well as to increased polarisation and decreased willingness to compromise on issues. Additionally, when we see shocking images or videos, our body may go into a 'fight-or-flight' response, which can trigger a physical and emotional reaction. This can include increased heart rate, sweating, feeling tense or anxious, or feeling overwhelmed. Our brain may also release stress hormones such as cortisol and adrenaline in response to the shock, which can affect our mood and cognitive functioning. Shocking images or videos that have gone viral because they trigger a strong emotional response, such as fear, disgust or anger, which can lead people to share the content with others in order to elicit a similar reaction and the rapid and widespread dissemination of information on social media, means that shocking content can quickly spread across the platform and reach a large audience in a short amount of time.[25] If you are interested in learning the culture of outrage, I highly recommend the academic paper 'Moral outrage in the digital age' by M. J. Crockett.

The visual is non-verbal and instead 'communicates' directly to our emotional brain and amygdala. And the technology then allows us to share that feeling, usually outrage, or fear, as an instinctive reaction. Even before we are able to verify if the image, video or story we may have got so upset about it.

Education and Experience: An Essential Combination

Let's have a look at the final element of this section. Alongside the largely unconscious learning process of conditioning, our conscious learning processes and life experiences also impact how we see and consume visuals, and the meaning we make out of them.

I'm grateful to my maternal grandparents for the out-of-school education they gave me while I was growing up. They'd often take me to their home where they would tell me stories and read to me, and we would often go on small trips to local scenic spots or museums. My grandparents also took me on a couple of cruises with them. I was probably about ten years old on our first trip together and I remember boarding this huge ship. The first stop on the cruise was Pompeii, an ancient city near Naples in Italy. I remember my horror the first time they told me the story of a town that was destroyed by the eruption of Mount Vesuvius in AD79. Preserved under volcanic ash, the town was excavated to offer a unique snapshot into Roman life. The eruption was so fierce and so sudden that as we walked through the town we could see people frozen in time. Pompeii is a UNESCO World Heritage Site and that year I was one of its 2.5 million yearly visitors.[26] While I felt that the house of my parents was restrictive visually, too tidied, limited any expansive thinking or visual joy, my grandparents kept opening the visual Pandora's Box with a world full of colours, different cultures and contrasts – one well worth living in.

When historic events are 'told' – verbally describing battles, victories, death and destruction – the telling does not quite conjure up the context in the multi-sensory and especially visual way witnessing the experience can. Remember that we are 65% visual thinkers and learners – visual in our absorption of information, context, and not so verbal. Thus, to me, it was just incredible to see history being so visually displayed. It helped me to understand that history had flesh, and texture. I could see what people had been doing just before they died, walk through the streets that they walked through and see the frescoes that they made. For someone with such a visual mind it was magical. Pompeii animated a lot of the dates and facts and figures from history books which had always seemed so abstract and yet, ironically, this felt so alive, vivid and real.

Being exposed to visuals from different parts of the world, and from the past, is an important part of our education. We see how other people live and expand our understanding of the world around us. When languages can be a barrier to exchange with different cultures, visuals aren't. My grandparents' introduction to other visual worlds opened new doors for me in my mind. And my sense of what was possible expanded with their dedication to my education and willingness to broaden my horizons. Their input also fired up my imagination. When we experience things visually, we learn about them on a deeper level. No matter how old you are, you can use this knowledge to seek out visual experiences that will help you better understand anything you are trying to learn or interpret. Museums are great resources for this kind of learning, and you can access a lot of their visual collections online even if the museum itself is not local. Even if you cannot change your visual environment right away, experiencing new visual landscapes, even temporarily, can change and expand your visual narrative.

Experiencing new visual landscapes helps give our imagination the tools to visualise new ideas. There's a brilliant book called *The Imagination Muscle: Where Good Ideas Come From (And How To Have More Of Them)* by Albert Read that specifically speaks on the power of imagination and being able to experience a scene visually.[27] I think that visualising more would enable us to comprehend fears better, to relate better to different types of people and events, whatever they are, including those who lived before us and those living today in different cultures.

Beyond experiences, our education impacts our ability to articulate, to yourself and to others, what it is that you see. This education often takes place in the classroom initially, but we can all build on our formal education at any point in our lives. Most people feel scared to vocalise their opinion on things that are artistic or visual because they believe they need and don't have the vocabulary to do so. We can say, 'I like it, I don't like it.' And that's basically it. If you don't get taught the subject, how do you speak about something like philosophy, in which we get to debate about what one opinion is for us versus another one. When we don't have a vocabulary to talk about visuals our arguments can fall flat very quickly because we can only talk about our personal tastes if we can't bring in language about different aspects of the visual and how it relates to a historical and social context. Widening the vocabulary we have to talk about visuals is important, which is one of the things this book is doing for you. After reading this book, you will ultimately better understand individual visuals and our shared visual landscape, which will help you play a more active role in shaping it. Having the language to talk about visuals opens up the art of visual critical thinking. This is the art of decoding a visual message, articulating our interpretation of it, and asking how or why it was constructed and shared with us. In the same way that we look at businesses, we need to break down visuals to understand the financial side of the business model that brought it to us. We need to comprehend what makes the visual message so that we can question whether we even want that visual message triggering us or whether we want to influence the way such visuals are made.

While it's true that all of us can tap into our imagination to inform our visual perception, often our formal education shapes our imagination and visual critical thinking in the first instance, and it is important to start there. I was lucky to receive all my schooling in France where private education isn't the norm. Everyone, from all different backgrounds and levels of wealth, is educated in the same school system and exposed to largely the same curriculum. This is not the case in the UK where wealthy children are often privately educated. Each school system, private and public, teaches a different curriculum and also creates 'bubbles' from which the people in each bubble experience life.

There is no denying that what we are exposed to as

children and through our education process will largely determine our relationship with our visual world. If we have access to private education, it often – but not always – means we were born into privilege. In those spaces, art is second nature, and there is usually rich and diverse access to all the arts. For those educated publicly, access to art is likelier to be down to each individual school and how determined arts teachers are to maintain vital access to these crucial subjects. But funding cuts often impact these subjects first.

My mum is a primary teacher in the public school system. I asked her – in order to determine how much time was spent on visual arts and understand the visual narrative surrounding us – to break down the day or week she had with her seven-year-old students. She's in France, so she led twelve hours of French in one week, five hours of maths, ninety minutes of English, three hours of sport, four hours of sciences, three hours each of history and geography, while only one hour was devoted to artistic or creative teaching. Yet we need to learn how to deal with those 10,000 commercial visuals each day! My mother's classroom is an example for me that – although we know that the majority of us will be visual thinkers – the school system is not in tune with visual thinking. The majority of us will benefit from improving our visual literacy outside of our formal educations.

This is just one example of a wider phenomenon. In 2018, a BBC survey suggested that 'of the state schools that responded, nine in every ten said they had cut back on lesson time, staff or facilities in at least one creative arts subject'.[28] Arts and culture writer Harry Hickmore reported on this phenomenon in 2019 in a piece for the *Huffington Post* entitled 'Cuts Mean Arts Education is Being Outsourced to the Culture Sector – And It's Not Working': his reporting revealed that 68% of primary schools in the UK had been experiencing cuts to their arts budgets in the past five years and music ceased to be taught in 50% of state-funded secondary schools.[29] This trend to cut funding in arts education has continued with Rishi Sunak calling for further cuts to higher eduction in the arts and humanities in 2023, only two years after former education secretary Gavin Williamson announced 50% cuts to arts courses.[30] I have always observed that art history is perceived to be a course that is a luxury, not like mathematics or science which are deemed essential. As an actor in the creative sector, I definitely have felt at times that my passion and now my job can be superficial or a luxury in the minds of others. It's never prioritised at school like mathematics would be and always seen as a little 'girly' (there was only one man in my art history class at Warwick University). I have always found it puzzling that we could have deemed visual literacy superficial whereas it is literally one of the most essential skillsets for navigating our overtly visual world.

In their book *Engines of Privilege*, authors Francis Green and David Kynaston state, 'In a society that mouths the virtues of equality of opportunity, of fairness and of social cohesion, the educational apartheid separating private schools from our state schools deploys our national educational resources unfairly and inefficiently; blocks social mobility; reproduces privilege down the generations; and underpins a damaging democratic deficit in our society.'[31] Reading it like this feels extreme and while I believe in the good intentions of our educational system, at least in our sector, I can witness the deep inequalities it creates every day in the endless amount of job applications that we receive that are disproportionately from privileged social economic backgrounds. We make an extra effort to reach out, offer training and venture beyond these backgrounds so that our team can be diverse, but it demonstrates that certain groups of people are more likely to see a career in the arts as a possibility for themselves.

This division means that often those from privilege (educationally, financially, or both) believe that art is part of their life. It also means that less than 2% of the working and middle class, according to the findings of the book *Culture is Bad for You*, go on to study arts at university.[32] As a talent agency owner, my mission is to democratise the very idea of art, away from the notion of dusty museums to the visuals we consume and interact with on a daily basis. More importantly my goal is to ensure that visual literacy shouldn't and doesn't come at a price. Everyone, regardless of background, can develop greater visual literacy which can in turn immeasurably increase the quality of our lives. Maria Balshaw, director of the Tate, one of the UK's leading art museums, is of the same mind. In 2019 she told the *Guardian* that 'Access to the visual arts in this country must not depend on social and economic advantage. Private schools place a premium on a rich cultural education for their pupils while many state schools are starved of the resources to support access to culture and creativity for their pupils.' She went on to add, 'We need a level cultural playing field for all children because we want and need visually literate adults. There should be fair access to arts in England in line with the offer to pupils in Scotland and Wales where the arts are already a core commitment.' Testimony from filmmaker and artist Steve McQueen highlights the need for democracy in arts education. He remembered his first trip to the Tate as an eye-opener to a concentration of visual creativity. His takeaway was that anything was possible. He said, 'The curriculum needs to be big enough to include all subjects and be for all children. Art and creativity are so important to science, to maths, or to any other academic venture . . .'[33]

Education is something that can equalise opportunities and access for people – although this is not entirely true due to a child's parents' backgrounds and influence. That is, if you remove education and leave it to the parents, you're just going into something that's lopsided and unequal. What we are exposed to via our formal schooling plus what we are exposed to in our homes creates expectations about what you learned.

What we are exposed to via our formal schooling creates expectations about what children from different backgrounds and different school systems can expect in terms of their life chances. Talent is all around us. Opportunity isn't always apparent.[34] As you read this book, our artist Rayvenn D. Clark has launched three large-scale sculptures in Alabama discussing the treatment of slaves across the US in the 19th century. At only twenty-seven years old, she will be in the 0.8% of financially successful visual artists.

Given her working-class background, she had less chance of accomplishing this than of winning the lottery (you officially have a one in 300 million chance of winning the lottery if we follow the latest numbers from the British press, so let's just say that Rayvenn really defied the odds!).[35] While Rayvenn is not alone in beating the odds, and many wonderful artists come from lower socioeconomic backgrounds, how much talent are we missing out on by not giving all children the same opportunities to succeed? We need more like Rayvenn in the visual sector so that we can insert new visual narratives into the streets we walk daily. Imagine the millions of people who will see her sculptures over the coming years, that's millions of people who will access a new visual narrative about slavery. It's history in the making with visuals. Who can say again that this is a superficial topic?

The Case for Visual Education

If I could effect a single policy, it would be to make visual literacy a compulsory life skill for all to learn within formal education. Currently, arts education is established internationally through a variety of mechanisms and structures, including national and international policies, funding, curriculum standards and teacher training programmes. The extent to which arts education is established and valued can vary widely across countries and regions, depending on factors such as cultural traditions, political priorities and available resources. Some countries have well-established arts education systems that are integrated into their national education policies and curriculums. For example, in many European countries like France, arts education is considered an essential part of a well-rounded education, and students may receive instruction in music, visual arts, drama and dance from a young age. In South Korea, there is a strong emphasis on arts education as a means of fostering creativity and innovation. Some African countries like Nigeria, Kenya and Rwanda and also South American countries like Chile, Colombia and Peru have made efforts to prioritise and strengthen arts education in recent years while, for example, Brazil and Argentina already have an established network of art schools and universities as well as initiatives that promote art education in schools. How would society look if we all developed our visual literacy skills? I believe it would reduce misinformation, improve communication and assist us in creating spaces that support our collective wellbeing. It would also reduce inequalities by accepting more voices into our overall narrative.

There is no denying that what we are exposed to as children and through our education process will largely determine our relationship with our visual world, but the good news is that we all keep learning throughout our lives. Even if you didn't learn about art in school, it's never too late to start. My goal is to ensure that visual literacy shouldn't and doesn't come at a price. Everyone, regardless of background, can develop greater visual literacy which can in turn immeasurably increase the quality of our lives.

Exercise
Local Learning

In terms of creative and visual education, the first thing that I would recommend is to go for a walk around a neighbourhood. You'll perhaps discover silly things: in a wealthy neighbourhood, you're likely to see very tall windows; you might realise that buildings have false columns that have no structural function but only an aesthetic one. Other areas might have large council flats, which are much more brutalist, or you might see a funny visual detail on your tube station stairwell.

Start by saying you don't know the visual narrative of the place you live in and go look for it. Find out what those visual references mean; is there a quirky story about any form of visuals that you're looking at?

Whether you start by going on a walk in your neighbourhood, or leafing through a magazine's adverts, just be open and learn about those visuals. You might find the way adverts are constructed really interesting – there are a lot of old masters' paintings referenced in the fashion magazines (can you spot them?). In your neighbourhood, learn about architecture, and about the history of the place.

Chapter 4.

The Anatomy of a Visual Message

'Naoshima Yellow Pumpkin' photograph by Kirill of a dock in
Naoshima, Japan, featuring *Yellow Pumpkin* sculpture by Yayoi Kusama,
© Kirill via Unsplash (25 October 2019)

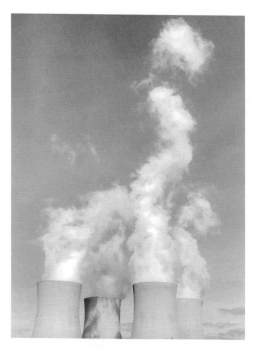

'White Power Plant' photograph of industrial smokestacks by
Jan Antonin Kolar, © JanKolar/Unsplash (25 September 2019)

'Sea' photograph of a wave by Yuriko David,
© YurikoDavid/iStock (30 August 2016)

Photograph of 'Project On Ice, Greek Sculpture' featuring the artist, David Popa. This artwork was created on ice floats in southern Finland using only earth, charcoal created from coconut shells and the source water. © David Popa

'Kim Marilyn' photograph of anonymous street art by
Aleks Marinkovic, © Aleks Marinkovic/Unsplash (23 May 2020)

Photograph of Choi Hung Estate, Hong Kong by Shark Ovski,
© Shark Ovski/Unsplash (7 August 2019)

'L.H.O.O.Q or La Joconde' altered postcard
by Marcel Duchamp, © Association Marcel
Duchamp/ADAGP, Paris/Artists Rights
Society/Norton Simon Museum (1919)

'Barbed Wire Sky' photograph by GluckKMB,
© GluckKMB/iStock (11 June 2014)

'A baby sitting on the ground' photography by Vikram Aditya,
© Vikram Aditya/Unsplash (13 September 2022)

'Madonna' digital art print by Àsìkò, © Àsìkò (2021)

Colour, focus, composition, perspectives and angles, distance and orientation. These are some of the components of what we see, and how these components are assembled can affect how we interpret what we see. These components make up a visual code that communicates meaning without saying a word. Unknown to us, these codes follow us as we browse fashion sites, watch the news, scroll on our phones or watch a film. They speak directly to a pre-verbal part of our brain – the area of our cerebral anatomy that gives us messages or feelings that we often can't even describe or articulate. These subliminal messages are stimuli, mainly audio and visual, that are presented in a way that the conscious mind cannot perceive or process them. They are below the threshold of conscious awareness, influencing our behaviour, habits, emotions and physical attributes without us being aware of what the messenger is doing. While subliminal messages may have no effect if they have no relevance to you, they can significantly influence you if they tap into an existing desire. For example, showing a subliminal message saying 'Drink Coca-Cola' won't make you thirsty. But if you are already thirsty and you see the same subliminal message, you are more likely to buy the suggested brand.

Take Mickey Mouse. On a global scale, it is probably the most popular cartoon character, and was imagined by

Disney employees John Hench and Marc Davis who believed that his circular head and ears would make him different to all other cartoon characters and more appealing to a broad range of audiences. Since his creation in 1928, he has become a symbol of American culture, as well as an international symbol of family-friendly entertainment. When you close your eyes and think of the Mickey Mouse image, you would imagine him as friendly, energetic, optimistic. He was given white gloves to contrast with his black body and help concentrate our attention on his expressive hand gestures, making them more theatrical. The illustrators picked red for his shorts to symbolise his energy and the sense of adventure and yellow for his shoes as they believed that it would make us feel positive when watching him. Mickey became the most marketable character of Disney, and the logo behind a family entertainment empire spanning several generations (to this day, Disneyworld and Disney Land Paris are the most profitable entertainment parks around the world, and have gained a cult-like following – the visuals created by Disney are consumed by millions of people of all ages every year). This careful plotting and symbolism turned Mickey into the most-famous rodent in the world (although, his girlfriend Minnie aside, recently Ratatouille has been vying for a runner-up title of sorts).

Beyond colour theory, which is exemplified through the above choices for Mickey's presentation, there are various fields of study that seek to provide a theoretical framework for examining how images convey meaning, including visual social semiotics, Gestalt theory, art history, psychoanalytical image analysis and iconography to name a few.[1] If you wish to take a deep-dive into the topic, I highly recommend Claire Harrison's article, 'Visual Social Semiotics: Understanding how still images make meaning'. The purpose of my book, however, is to give a general entry-point into the inner workings of the visual narrative. I want everyone to understand that everything, right down to the colour of your shoes, was chosen on purpose as part of one of the most lucrative marketing campaigns of all times. In this chapter we will explore the visual codes that images transmit through various characteristics from colour, to size, star (or focus), composition, orientation, props, source, intention and veracity. It's important for you to understand how these decisions are made so you can better interpret, not just Mickey Mouse, but everything around you.

'Bigger is Better'

Instinctively, we all believe that size matters. Just like how we have already discussed the fact that the largest billboards in the London underground are synced to match where people stand the most.

Upon first seeing Leonardo da Vinci's *Mona Lisa*, or *La Joconde* as it is known at the Louvre, 'the best known, the most visited, the most written about, the most sung about, [and] the most parodied work of art in the world', many people express surprise at its size, which at 77 x 53cm feels like something that could be in their own homes.[2] This is somewhat understandable given that many of the most famous paintings in the world are very large, at least when compared to the kind of art most people have on their walls. Think of Monet's *Waterlilies* series (if placed in a row they would span almost 91 metres), Botticelli's *The Birth of Venus* (1.72 x 2.78 metres) or Raphael's *Sistine Madonna* (2.69 x 2.01 metres). They are magnificent, awe-inspiring declarations of the significance of the subject matter, the wealth and status of the patron and the sheer craftsmanship of the artist, and over the decades their great scale has, in our perception, become intertwined with their great fame. Consequently, we expect all the famous paintings that we have seen countless times on posters and merchandise, in books or in movies, to be as grand in size as they are in prominence. So it can come almost as a shock when we see a painting like Vermeer's *Girl with a Pearl Earring* in real life and realise that it is in fact a mere 45 x 40 cm.

There's a reason why we make these assumptions. Have you read the small print on your mobile phone contracts or the terms of service for social media platforms? If so, you're more informed than average – most of us don't do it because the size of the information makes it unappealing, as does the length of the transcripts and requisite legalese. The same is true of visuals – large is a foghorn to the whisper of small. Perhaps we should all pay a little more attention to the whisper.

Some questions to consider asking around size when you encounter a visual:

- Does the size of this visual or imagery tell me something extra about the visual and its importance? Is that extra message warranted or contrived?

- Is there a size contrast in the image? If so, which elements are shown as larger than others or smaller than others. Again, what do you think that contrast is seeking to tell us? Is that accurate or manufactured?
- Is it possible that the small details are an attempt to hide in plain sight? If so, why?

The Power of Colour

Thousands of years ago, early humans in Northern Africa first realised that they could use pigments to communicate. The very first 'man-made colour' came from charcoal or earth pigments such as limonite, hematite, red ochre, yellow ochre, umber, burnt bones and white calcite, which were then ground up into a paste and mixed with binders of spit, blood, urine, vegetable juices, or animal fat. These colours were used to decorate their bodies and objects around them, symbolically telling each other who they were and information about the world around them, from practical information about where to find the best food to religious images of gods and spirits. The ability to create art using colour spread all over the world, changing with the people who made it wherever they went.[3]

Our perception of colour is affected by our gender, age, culture and ethnic background, among other factors, so we all experience colour differently, but taking a step back to examine how colour is commonly understood and used to communciate can help us better decode the visuals around us. For example, in modern Western cultures white is often used to represent purity, cleanliness and neutrality, and yet it is a symbol of mourning in many eastern cultures. During a Hindu funeral, mourners will wear white to show respect to the departed and the family. It would also be common for the deceased to be wearing white clothes when buried.[4] However, there are also certain similar universal reactions to colour that transcend the individual variables.

Colour plays a significant role in not only what we pay attention to in our visual landscape but also how we interpret those colours subconsciously. Brand experts know this and will use these insights to create brands that attract our attention and encourage the response that is required – through colour and design. Research has shown that 93% of customers examine visual appearance when making a purchasing choice.[5] This is obvious to some extent but the influence of visual appearance on our purchasing decisions is a lot higher than we might imagine! Another study has shown that most customers make a subconscious assessment about which brands or products to choose within ninety seconds of an initial viewing and up to 90% of that assessment is based on colour alone.[6]

Let's take a look at some common colours and their corresponding psychological values:

Primary Colours

We call just three colours 'primary' – red, blue, yellow. They are the basis for all other colours that we create. Mixing two primaries creates a 'secondary' colour. In school, very young children learn that mixing two secondary colours creates yet another colour that we classify as 'tertiary'.

Yellow

When I think of yellow, it's impossible for me not to think of Van Gogh's *Sunflowers*. On the 73 x 91cm painting, the entire canvas is made of different shades of yellow, from a mustard yellow up to lemon yellow. Paul Gauguin, the painter and friend of Van Gogh, wrote about his use of yellow: 'Oh yes, he loved yellow, this good Vincent, the painter from Holland – those glimmers of sunlight rekindled his soul, that abhorred the fog, that needed the warmth.'[7] Yellow is seen as a strong, attention-grabbing colour that invokes optimism, sun-like warmth, youthfulness and energy. Stuck by many dark obsessions caused by depression, yellow was meant to illuminate Van Gogh's days and exudes his wish for hope, happiness and childlike simplicity. This connection between yellow and optimism and joy is one of the reasons that Disney chose a sunshine yellow for Mickey Mouse's shoes and why McDonald's chose yellow for their famous golden arches.

And yet, in Roman times, 'seeing yellow' was described by Lucretius, Varro and Cassius as a stigma of the mentally unsound (and they were specifically looking at mental health).[8] Therefore, the Romans would have read the meaning of the yellow sunflowers of Van Gogh very differently to us. Context is invaluable when looking at visuals from a different time. The meaning of colours keeps evolving with our societies.

Red

Our second primary colour, red, can be triggering and often signals alarm (hence stop lights). This is evidenced by its prominence in the horror genre (think of *The Shining*'s 'redrum', or Hitchcock's psychological thriller, *Marnie*). Personally, my worst association with red was during the Brexit campaign when multiple red buses were going all around England making a statement that has since been disproven: 'We send the EU £350 million a week, let's fund our NHS instead. Vote Leave.'[9] Of course, even close to eight years after the fact, the NHS didn't get any extra funding – in fact, currently doctors and nurses are on strike as their working conditions worsened after 2016, the year of the Brexit vote. This red bus felt like a nightmare, it was spotted everywhere but more importantly: it was effective, it helped win the referendum.

Red is a warm colour but its meaning can be contradictory. On one hand it is considered the colour of love and passion but it is also considered the colour of anger and power. This is another colour that means very different things in different places across the globe. In UK politics, red represents Labour or left-wing politics, whereas in the US red is the colour of the right-wing Republicans. Red is also a popular colour in Chinese culture, symbolising luck, joy and happiness, and also represents celebration, vitality and fertility in traditional Chinese colour symbolism – it is also the traditional colour worn by Chinese brides, as it is believed to ward off evil.[10]

In her show titled 'The Artist is Present' at MOMA in 2010, performance artist Marina Abramovic wore a red gown to convey her ideas silently to her audience. Over the course of nearly three months, for eight hours a day, Abramovic met the gaze of 1,000 strangers, many of whom were moved to tears while facing her silently in her red dress.[11] The dress colour communicated something to the audience and in turn made them feel a certain way, it reflected their own emotions. Now imagine the cultural differences between each audience member: they would each perceive a radically different meaning from this dress. The omnipresent silence made their interpretation the absolute truth.

Blue

Our final primary colour, blue, is a colour (along with green) that is seen most often in nature, certainly in the northern hemisphere, and so it often elicits a sense of serenity, peace and calm. It is used in corporate livery to denote reliability, trust and order. Britain's National Health Service's (NHS) livery and the European Union flags are blue. I specifically chose blue for the colour of the logo of my company MTArt Agency.

In creating the colour scheme for my agency (and without overthinking the matter too much), blue came to mind for several reasons. Blue represents loyalty, trust and longevity. Infinity is loyalty and so is related to longevity. I wanted my firm to last. I wanted people to relate to and trust us. A deep navy blue feels very conservative and is described as very traditionally minded. And because we wanted to disrupt the way the sector was going, a turquoise colour felt more active or energetic and therefore more aligned with our intention – although we cared about the long term and the possibilities, we also cared about the energy we were putting in and the fact that it was not traditionally minded.

Blue is also connected with sadness. This can be seen in Pablo Picasso's paintings during his 'Blue Period'.[12] Blue in these paintings is cold and deep, signifying misery and despair. It highlights the hopelessness of the figures depicted, such as beggars, prostitutes, the blind, out-of-work actors and circus folk, as well as Picasso himself and his penniless friends. At the time, he only wore blue clothes. Sadness, melancholy and reflection are also omnipresent in the pale blues of 2021's 'Reverse Selfie' Dove advert.[13] This advert was designed to show the pressure and negative impact social media is having on young girls in particular. Starting with a selfie of an arresting woman, the selfie is reversed, showing the makeup being removed and the filters being removed until you see who that 'woman' really is – a very young girl. The girl, who has been rewarded with likes and comments about how amazing she looks still appears sad because she knows the image is not her and all it's done has been to make her feel even more insecure about how she looks. The girl's expression is neutral throughout the clip, but the blue tint over the whole video provides the visual cue we need to interpret the emotional message.

Secondary Colours

Green

Green is associated with the environment and nature more than any other colour, a link that is so strong it is often used deliberately to make something seem environmentally friendly when it isn't. At the time of writing this, there are two lawyers trying to get a law passed in France which will be called 'the Right to Water'. If successfully passed, it will apply to our two famous French water companies: Evian and Vittel. Today, people are no longer allowed to build houses near the Evian and Vittel factories because there isn't enough water for those who live there. This is because the two water companies have used the water deep down in the aquifer at a faster rate than it can naturally be replenished. Now, we like to think of water as something that is universal and accessible to everyone, but in this case it is being taken by a private company and sold for profit.

A keen observer will notice that Evian and Vittel are specifically using the same green colour in their adverts. Both use green to show how natural and how trustworthy and how aligned they are as brands consumers can trust. It is a literal example of the term 'greenwashing' – the use of the colour green to make something eco-friendly and natural by tapping into our strong association between the colour and nature.

Michel Pastoureau, a French academic, recently ran a survey on the colour green in 2018.[14] Of the people he interviewed, 20% said that this was their favourite colour, and yet another 20% said that this was the colour they hated the most. Puzzled, Pastoureau went to study why green can be so polarising. In the arts, it's known that actors categorically refuse to wear the colour green on stage – this comes from the time when Shakespearean actors wore a special tint of green on their clothing, made out of toxic pigments from the oxidation of copper.[15] Copper reacts with the oxygen in the air to form copper oxide which creates this iconic blue-green patina colour which, while striking aesthetically, is highly toxic.[16] An inconceivable number of actors died on-stage from breathing this tincture. Some also say the French dramaturg Molière died on stage while wearing a radiant green costume.[17] This only reinforced a dislike of the colour green, and occasionally, in its more fluorescent iteration, the colour is used to represent slime and magic (e.g. the musical *Wicked*).

But green also has positive associations: deep greens are associated with nature, healing, safety and luck – and are also used to evoke calm or compassion. Learning about the history of the colour green can demonstrate how cultural associations with colours evolve over time. Historically, the earliest use of green for nationalistic reasons in Ireland was seen during the Great Irish Rebellion of 1641, in which displaced Catholic landowners and bishops rebelled against the authority of the English crown.[18] It cropped up again in the 1790s to bring Republican ideas to Ireland, inspired by the French Revolution. The main group that promoted this idea, the Society of United Irishmen, wore green. The continued importance of the colour green for Irish people all over the world is especially visible on St Patrick's Day. My husband is an Irishman, so green is everywhere in our household, for us to remain lucky.

We can see the health connotations of green in the late Queen of England's 'We Will Meet Again' speech, broadcast from Windsor Castle during the 2020 Covid lockdown. The broadcast was filled with green imagery, including in the Queen's own regalia.[19] Green here was used to express reassurance and hope to the English people. The Queen's choice of colour here was not random, she was known for using the symbolic power of colours throughout her entire reign. In her book *Our Rainbow Queen: a Celebration of our Beloved and Longest-Reigning Monarch*, Sali Hughes deconstructs how she used colours to send strong statements in all her appearances. Each day she doubled down on one colour, wearing it head to toe. While on royal tours, she also made sure to incorporate the colours of the national emblems she visited to communicate connection, as visible in all the photographs of her public appearances from all over the world.[20]

Orange

When I think of the colour orange, I always think of the painting *The Scream* by Edvard Munch and the boiling sky behind the character in the forefront. This colour was inspired by a particularly strong sunset caused by the Krakatoa eruption, one of the most dramatic volcanic eruptions ever witnessed in the 20th century. Orange is assertive. It is often used to demand attention. It can also confer excitement, warmth and spontaneity. We can find it in the spicy pumpkin meals of autumn. Orange is loud. Too much orange can be overwhelming and used as almost self-aggrandising. In Ukraine, the colour orange means strength and bravery, whereas in the United States, a darker shade of orange is almost retributive as it is used in prison uniforms, associated with untrustworthiness and deceit. An old saying that my granny passed on to me is that it stimulates appetite. It can trigger hunger in its reddish hues and it is often used in kitchen decor and restaurants for that reason. If you have children who don't eat as much as you would like to, try adding some orange details to your kitchen or dining room. That's perhaps why worldwide, beyond orange as a fruit, orange the colour is associated with vitamin C and healthy eating.

Purple

If you are plugged into British politics, you will remember the purple and yellow logo and the lion of the nationalist party UKIP during the Brexit campaign. UKIP's yellow and purple are supposed to evoke prestige and be reminiscent of white British heritage.[21]

Purple is also often associated with royalty and power because historically purple dye used in ancient times was very rare and extremely expensive. The resources needed to create a dye in this colour were much harder to come by (since purple is uncommon in nature) and much more costly. So only the elite could access purple dye. Around 1200 BCE, the city of Tyre (along the coast of ancient

Phoenicia) began producing purple dye by crushing the shells of a small sea snail.[22] The resulting colour became known as Tyrian purple and was so well known that it was mentioned in Homer's *Iliad* and Virgil's *Aeneid*. Alexander the Great and the kings of Egypt also wore clothing coloured with the famous Tyrian purple. This connection between purple and royalty has continued to modern times. It was used by the musical artist Prince to create a regal persona. Purple was also the colour of choice for the robe of estate worn by Queen Elizabeth II on her way back to Buckingham Palace following her coronation in 1953.[23] More recently and within the LGBTQ+ community, the purple of the pride flag represents non-binary gender identities. In the bisexual flag, the pink and blue overlap to form purple representing bisexuality.

Pink

Pink is often used to indicate that a product is 'for women' or 'for girls' because of its strong association with femininity. The phrase 'pink collar' still designates jobs held by women, like sales clerks and secretaries, caregivers and other professions seen as traditionally for women.[24] Audrey Gelman and Lauren Kassan knew this well, and used pink to create an instantly recognisable brand for their feminist co-working space The Wing.[25] Over the course of four years, their start-up grew rapidly, operating in twenty spaces across the USA and in London where the most prominent feminist writers, artists and media personalities mingled.[26] The start-up ultimately closed in 2020, citing financial impacts during the Covid pandemic. Throughout its existence, it triggered a lot of reactions online and in the media. In its early years, it was featured in *Vogue* and *Forbes* as an up-and-coming space to empower women professionally.[27] Podcast host Aminatou Sow said, 'I see more women breastfeeding there than I've seen anywhere else,' indicating that it did succeed in creating a new type of workspace for women; however, it also received a lot of criticism for its price point, allegations of racism and its focus on women in business, which pulled it into a complicated ongoing conversation about how capitalism coopts feminism for profit.[28] The *New York Times* writer Amanda Hess called it 'Instagram-ready feminism'.[29] From girlboss feminism to pink razors, many have come to look at pink cynically when it appears in branding and marketing.

However, I cannot mention pink in 2023 without talking about the surge in people embracing all things pink sparked by excitement for Greta Gerwig's *Barbie* movie. The film has grossed over $1 billion as I write this at the end of summer 2023, making Gerwig the highest-netting female director of all time.[30] The Barbie doll is the ultimate pink icon, she represents an unabashedly 'girly' femininity, and this outpouring of enthusiasm for pink feels like a post-ironic reclaiming of all of its joyful connotations and connections to girlhood. We bring a lot of our feelings about feminity to how we interpret pink.

And yet pink wasn't always seen as a feminine colour. It was once the favourite colour of male aristocrats in Europe in the 17th and 18th centuries.[31] Pink only became associated with femininity after World War One. A group of marketing strategists, clothing manufacturers and retailers worked out that they could sell double the amount of clothing if they could establish that one colour was associated with one gender. Initially, blue was associated with girls and pink with boys, but by the 1940s this association was reversed.[32]

We can also think about pink through the lens of its parent colours. It is a mix of red and white. The red undertones of pink can give it a sultry or romantic connotation. The white elements found within pink counteract this with its purity. The deeper the shade is, the more passion it exudes. Meanwhile, paler tones give off a more virtuous vibe.[33]

On the negative side pink can be viewed as lacking willpower, self-reliance or self-worth. Too emotional and too cautious just like piglet in *Winnie the Pooh* whose insecurities we remember from our childhood.

Brown

The *Mona Lisa* by Leonardo da Vinci is interesting in a discussion about the colour brown, but not the first example most would name. The use of brown tones is pervasive, yet we never think of the portrait in terms of colour choice. We always think of her gaze and her enigmatic smile. The browns escape us here, perhaps because they are solid, timeless and almost kind of boring.

Brown tends to feel like a solid, earthy colour, but it can sometimes seem drab and boring. Positive qualities of brown include wholesome, earthy, natural, reliable, healing, warmth and honesty. Think of a robust and tall oak tree. It can also be considered luxurious or elegant. It is often used in corporate livery to denote the quality of being down-to-earth and that's why so many corporations have brown in their logos: UPS, JP Morgan, M&Ms, Hershey's, etc. In feng shui, brown represents either wood, if it's dark and rich, or earth if it's light and should be well balanced with other colours to avoid a lack of ambition and drive.[34] Psychologists have found brown to increase the levels of tryptophan in the body, which helps our sleep and supports our immune system. It is also thought to increase our serotonin levels, an essential chemical for mood regulation. Brown pigments are among the oldest and were often used in prehistoric art. In Eastern and Asian cultures, brown is the colour of mourning.[35] In many countries brown eggs are believed to be more nutritious and healthy and natural, even though there is no scientific basis to back this up.[36] Free-range and organic eggs tend to be brown too, which reinforces this opinion. After the psychedelic sixties, brown became a very popular colour in the seventies, maybe also reflecting increasing environmental awareness marked, for example, by brown being the main colour used in branding the first Earth Day celebrated on 22 April, 1970.[37]

It can evoke a sense of warmth, comfort and stability, as well as security and groundedness, and is often associated with autumn. Brown is commonly used to represent reliability, simplicity and understated elegance. It can also be seen as rustic, vintage or traditional, and can evoke a sense of nostalgia or comfort.[38] While brown may be useful for a rugged appeal – for example for leather products – when positioned in another context, brown can be used to create a warm, inviting feeling (Thanksgiving) or to whet your appetite (every chocolate commercial you've ever seen).[39]

White

In Yasmina Reza's play *Art*, a contemporary art collector buys a painting that has layers of white paint for a lot of money and introduces the artwork to his friends.[40] This sparks one of the strongest debates about contemporary art: its lack of meaning, its exorbitant value and its inaccessibility. As they keep arguing, they keep staring at the white painting. Here, the white painting, in its purported neutrality, feels like a deep silence in front of the superficial noise that our society can create. White, which can evoke blankness and empty space, can be arresting and eye-catching because we are primed by our repeated experience with white paper to expect it to be a canvas for a message.

One of the best-known uses of white is in wedding dresses, where it symbolises purity. While wedding dresses are traditionally white in the Western world, this was not always the case. Up until the 1800s European cultures also associated white with mourning.[41] It was Queen Victoria who introduced the tradition of the white wedding gown, when she wore a relatively simple white dress, constructed of British-made materials, especially lace from Devonshire to support the struggling lace industry, which she felt would be best highlighted on a white dress.[42] The white wedding dress soon became widely popular.

White has been positioned to conjure a sense of purity and innocence in Western culture. But it can also symbolise sadness, death, mourning and peace, as we saw with Hindu mourners wearing white. At the height of the Covid pandemic, *Vogue Italia*'s April 2020 edition featured a blank white cover. The starkness of the white cover signaled a time of reflection. As the virus continued to devastate Italy, the publication took it seriously. 'To speak of anything else – while people are dying, doctors and nurses are risking their lives and the world is changing forever – is not the DNA of *Vogue Italia*,' wrote editor-in-chief Emanuele Farneti.[43] This reminds me so much of the power of a white flag and its silencing presence – all the while reminding us of what matters: peace, and in this case, health.

White is also considered a cold, bland colour. When I was growing up, it was a stark reminder of the cleaning obsession of my mother, it felt like I was living in a sterile hospital.

Black

I was sitting on a café terrace with the artist Elisa Insua and we were discussing the fascination that people had for the character Batman. His origin story sees him swearing vengeance against criminals after witnessing the murder of his parents Thomas and Martha as a child, a vendetta tempered with the ideal of justice. Batman embodies the symbolism of the colour.[44] Black here is associated with power, fear, mystery, strength, authority, elegance, formality, death, evil, aggression, rebellion and sophistication.

This polarisation is best perceived if we compare the artworks of John Singer Sargent's *Madame X* with Fransisco Goya's 'black paintings'. The former is a portrait of

Parisian socialite Virginie Amélie Avegno Gautreau, who was renowned for her beauty and infamous for her affairs, her stunning black dress in the picture an emblem of elegance, sexiness and provocation. The latter is a series of macabre, haunting images featuring death, insanity and witchcraft, including the painting *Saturn Devouring His Son*, probably inspired by the artist surviving two near-fatal illnesses and witnessing the terror and turmoil of the 'liberation war' following Napoleon's occupation of Spain.

Black is also required for all other colours to have depth and variation of hue.

We see the depth of the prejudice of the colour black also reflected in language: black magic, black hearted, black sheep, blackmail, black market, etc. The bad guys in movies are usually dressed in black to create mystery around the character's identity. Darth Vader wouldn't have had quite the same impact if he'd worn pink!

Metallics

Bronze

The well-known statue, *The Thinker* by Auguste Rodin, is the first bronze sculpture that really got me to stop, look and think when I was younger. The bronze gives it its timeless nature as he thinks forever – much like the timeless philosopher most people associate him with. The heroic-sized nude body is already powerful, evocative, but the bronze material adds to the weightiness of his thought, the timelessness of human thought.

Bronze is closely related to brown, but its shine gives it a more luxurious connotation. It is mostly viewed as a winter colour that often depicts endings, such as dying leaves in autumn or the setting sun – it is a comforting colour. It is also the colour of third-place medals in the Olympics – symbolic of a great achievement and hard work. People associate it, as with *The Thinker*, with strength, stability and support. It's often said that if bronze was a person, it would have a great head on its shoulders. In the Bible, bronze is a symbol of God's righteous judgement due as a result of sin. As you walk around a city, public sculptures are often made of bronze and it almost feels as if the bronze gives them their timeless and wise appearance.

Silver

Around 70% of all motor vehicles are silver, grey, white or black, with grey and silver at 28% of the whole.[45] Back in the days of very early automobile production, black was a much cheaper and easier colour to produce than silver, giving the more expensive silver vehicle a patina of wealth and uniqueness. Silver (along with these other common colours) vehicles retain the greatest residual value because they are easiest to resell.[46]

Silver, as a precious metal used for jewellery and expensive objects such as tableware, is the colour of affluence and modernism. It is both healing and dynamic. The colour silver is associated with industrial, sleek, high-tech and modern, as well as ornate, glamourous, graceful, sophisticated and elegant. In the 19th century in Europe, it was also believed that silver would draw negative energy out of the body and replace it with positive energy. Traditionally, grey-haired older people are viewed as just being old, while the phrase silver-haired traditionally describes a distinguished individual who is ageing gracefully. The saying 'silver-tongued' is used to describe a witty and eloquent speaker, while the expression 'silver-tongued devil' refers to an articulate speaker who is insincere and possibly a liar.

Gold

When I think of gold, I always think of Beyoncé, who asserted her pop queen status in gilded clothes. Looking up her name and the word 'gold', the golden outfits she wore in her media appearances and her music clips are endless.

Gold spells opulence. Unsurprisingly, it is the colour of wealth, success, knowledge and confidence. When establishing his power across Europe, Louis XIV filled his Chateau de Versailles with gold, which was crucial to enhancing the king's image as the self-professed sun king – an apollo bringing light to an unenlightened world. To give you an idea of how much gold the palace holds, it took 100,000 gold leaves to cover the 80-metre steel gate.[47]

Throughout history, gold has been used to reinforce a very symbolic meaning. The Christian church commissioned many golden artefacts from the 15th century up to the 19th century to introduce its saints and icons to believers in its various churches across the globe. Hollywood did the same with its Walk of Fame, a historic landmark which consists of more than 2,700 gold stars with the names of the most famous people in the world of cinema along fifteen blocks of Hollywood Boulevard. More recently, when we walk around Dubai and Doha, gold is everywhere, from the mosques to grand offices as they established their new domination geopolitically. Gold has fascinated us for centuries – think of the many appellations associated with the colour: gold digger, good as gold, gold star, gold standard, a heart of gold, golden ticket, etc.

Colour Checklist

Having read this section, you might find yourself questioning the colours chosen in the art, clothing, ads or even interiors around you and what impact those colours have on how you feel and how you perceive the visual message. For example, a study has found that warm-coloured placebo pills are more effective than cool-coloured placebo pills.[48] The warm-hued pills tapped into the optimistic feeling we get with warm colours. And there is also anecdotal evidence in this study that blue-coloured street lights reduce crime, a reflection of the soothing and calming properties of blue. Other aspects of colour such as tone and contrast also have an impact, for example higher-contrast images tend to be more attention-grabbing and energising.

When looking at visuals, it might be helpful to recognise that colour is playing a subliminal role in our attention to and interpretation of various visuals. Questions you may find useful include:

- What is my immediate gut instinct to this visual? In other words, how do I feel when I look at the image before I start to analyse it?
- How does colour influence it?
- What is the meaning of the colours used?

Whether you are choosing an outfit to wear, designing artwork, picking out 'save-the-date' designs, or indeed making sense of a few different dynamics within your workplace or elsewhere in your life, these tools will help you to operate with the knowledge that guided some of history's top merchants of culture. Turn to page 203 for guidance on how to apply some of these insights in your own space.

Focus: The Centre of Attention

Artists and advertisers use deliberate techniques to help draw your eye to a specific 'star' or focus within an image. The message we take from a visual is shaped by how the star of an image is portrayed or where the focus on the visual is placed.

For example, in *The Last Supper*, our eyes are naturally drawn to the central point of the picture: the 'star' is Jesus surrounded by his twelve apostles. This is only one of countless paintings of Christ, and often these images feature him right in the centre. In this painting, depicting a story told in the Gospel of John, 13:21, Jesus has just announced to his apostles that one of them will be betraying him, so Da Vinci portrays all the reactions rippling off that announcement. Every apostle on either side ultimately has a different reaction to hearing that news, but ultimately their reactions all help draw the eye back towards Jesus, the centre of the image and the main character in the message being conveyed by it. By making Jesus the main focus of the image, Da Vinci positions him clearly as the hero of the story, drawing us, via our attention, to identify and admire him as the central figure while seeing the other characters in the image as secondary elements to the story.

Humans have a need for heroes, and heroes are one way we shape narrative in images. Today's visuals are being shaped in a similar way. Take Beyoncé. Whether on stage or in portraits, she is usually depicted taking that centre stage as a hero with supporting dancers or visual elements drawing the eye towards her. When you look at an image, pay attention to where your eye is drawn to first – this is likely the key to the message the visual is trying to send. In *The Last Supper*, every element of the image is drawing your eye towards Jesus.

The images and subjects used in a visual are important because they indicate to us who it is speaking to and reflect to us an image of how the world is, or could be. So it is worth paying attention to who or what is represented in a visual because it tells you something about the worldview expressed by the image. For example, if you were to visit any major historical museum anywhere in the world you would be hard pressed to find any stars (or heroes) in any paintings that are not white, usually male. This is largely because any people of colour, if depicted at all, are depicted as servants or worse. This is because for a long time Europe was all-Caucasian and males were the only accepted decision-makers of society.

Lynda Mead argues in her book *The Female Nude: Art, Obscenity and Sexuality*, that if you observe art history over time, you see that female roles change when comparing art from the Middle Ages, where under the feudal system women and men were mostly equal, to the Renaissance, when paintings primarily portrayed women in two ways, as either virtuous and chaste (or motherly) or seductive and deceptive.[49] Women have almost always been depicted in the background, as subservient, and have been restricted to a few very specific roles, whether in real life or in how they were represented. I think this can be explained in terms of the fact that action was something that was thought to be a male domain historically, while the act of being observant, visually pleasing – visually observing objects and events – was more a feminine trait. John Berger famously wrote in *Ways of Seeing*[50] that 'Men look at women. Women watch themselves being looked at,' to describe this idea that the expected viewer of an artwork in a European painting was a man, which, like depictions of people of colour, echoed real-life power dynamics in society. The images were both shaped by culture and became shapers of culture, reinforcing the dynamics they depicted through the visual message embedded within them.

Though we've come a long way since Berger was writing in 1972, we see similar ideas about how we depict people of different genders continue to pop up. For example, in the top 100 grossing films of 2019, women were more likely to be cast in supporting roles than lead roles.[51] Thankfully this is changing with female directors such as Greta Gerwig breaking box-office records by telling stories that centre women. In the early 1900s, Sigmund Freud coined the term 'the Madonna-Whore complex', identifying that many men see women as either pure, chaste and maternal Madonnas or sexual objects.[52] Freud coined this as a psychological phenomenon, but I see it as a visual problem. This view of what women can be has long spilled into the visuals that we create. Because art history has mainly recorded visuals of Madonnas or whores, we couldn't imagine women to be anything else, but when we consume visuals that counter and complicate this binary we can break down the internalised bias.

Focus Checklist

We'll always want to have people we look up to. Sometimes it's that whoever seems to be at the top of our society is who we will strive to emulate. Heroes must be the right visual nowadays, however, making sure that we don't accept a hero or star just because we've been so used to seeing them. We should always challenge them, and even, perhaps, look for a visual that's uncomfortable or that is very different to what we have in place. This is becoming aware of our narrative, by constantly challenging the way we are looking at things.

It obviously will influence us constantly because we are led to think maybe we're doing everything wrong unless we are the type of star or hero in the centre of the action, such as Beyoncé has become. Visuals of the past ultimately shaped who we wanted to become and we can change them. If we do not consciously challenge our visual landscape, where we erect demi-gods, which is Hollywood ultimately, we end up always measuring ourselves against them.

The questions to consider when assessing the imagery that is shaping our visual narrative – stars and heroes and the whole of our visual world and lives – include:

- Who is the represented participant in the image? This can be either a person or an object or an element from nature. What is the focus of the image? What does the choice of focus tell you? Is that choice of focus likely to be deliberate and if so, what message is that choice probably seeking to elicit from you? Just be aware of the potential influence of that choice.
- Is there anything about the image or visual that indicates action? If so, what kind of story does this action infer?
- Are the people in the visual looking at each other or away from each other? What does their proximity to each other and visual connection or lack of it tell you about their history?
- Is the choice of star or focus trying to tell you anything socially or culturally?
- Why that star or that focus? What message is it attempting to send? Is it accurate? Is it truthful? Is it fair? Is there anything else in the visual that is reinforcing that central idea?
- Is the image or visual confronting you?

Composition

As well as the choice of star or focus, how that visual is then presented in terms of composition also sends intentional messages that we may not be consciously aware of. A visual can fail or succeed based on its composition and how well it leads the eye to the focal point.

The famous 'tank man' photograph by Jeff Widener from the Tiananmen Square protests in 1989 is an extremely powerful image in terms of composition. The way the tanks line up diagonally across the picture, the gun barrels pointing towards that proportionally little man in front of them, really draws the viewer's eye to him and makes him the focal point of the image, even though he is the smallest object and not even in the centre of the picture. The composition also adds to the sense of isolation and vulnerability of the man, as he appears small and alone in comparison to the massive tanks. I would argue that the photo's composition helped to make it the iconic image of peaceful resistance and symbol of democracy and human rights that it is.

There are several features that are used in the visual landscape to elicit a feeling or demand attention or action. Being more aware of what they are should allow you to spot these tactics and disengage from them.

In her paper, 'Visual Social Semiotics: Understanding How Still Images Make Meaning', Claire Harrison posited that these features include:

- Image action and gaze
- Social distance and intimacy
- Perspective – horizontal angle and involvement
- Perspective – vertical angle and power

I think that Claire's framework is helpful because it's easier to understand a visual when breaking it down rather than trying to understand it all at once. Focusing the eyes on specifics further helps the reading of the image. I've used her general framework here to help guide us through the subtler elements that we need to consider while analysing a visual message.

The Gaze

If I close my eyes right now, I can recall the smily eyes of my son as he left for school this morning with his small car in his hands or the deep gaze of Granny yesterday when I left her after a long conversation. These images stay imprinted in my mind for years. Most of us unconsciously take in a lot of information from a person's body language, and our eyes are one of the most expressive parts of our bodies. We unconsciously look at the eyes and follow a person's gaze to understand them. The gaze is really interesting because we use it ourselves in our daily lives to create a visual message. For example, a listener might raise an eyebrow at a speaker to convey scepticism and we frown directly at someone to convey displeasure. In art, media and film the gaze is both who we think is looking at us (and how we interpret the meaning of that particular look) and our point of view as directed towards the image, which was intentionally created for us to absorb as our own. This point of view that we inhabit when gazing at an image tells us about the image creator's visual narrative and conveys ideas that we feel reflected back on ourselves. In other words, it can be both what I imagined that I would be seeing if I was looking at myself, and projecting a gaze from someone else onto myself.

The only theory by another that I will present here is dubbed 'The Male Gaze' theory. It is the work of Laura Mulvey, a film theorist mostly known for her comments regarding sexual objectification of women in media.[53] This theory has been used to identify issues with gender in film. Her position was that the male viewer is the target audience, and thus his needs and interests are met first in the construction of the visual and that this circumstance stems from an old-fashioned, male-driven society. Mulvey writes that even a female photographer taking a photo of a woman is influenced by the male gaze because it is so pervasive in our culture. There are other gazes that shape our culture too, such as the 'white gaze' coined by Toni Morrison, which refers to the assumed white reader or viewer of an image.[54] The dominant gaze within our culture is shaped by who has traditionally wielded the most social capital and power.

We are constantly pressurised by the gaze. And we're born with that gaze from the minute we're born, with people projecting onto us who we should be. This should lead you to keep in mind that there are different types of pressure and meanings associated with a gaze. I believe that the key to combatting the gaze is to increase our awareness of our own biases and challenge it directly by bringing more varied perspectives to the images that make up our visual landscape. When viewing an image, think about who the intended audience was. How does your interpretation of an image's message change when you put yourself in someone else's shoes?

When we're looking at an image, we also need to consider the 'gaze' of the image's subjects too. Journalist Heidi Mitchell recently reported in the *Wall Street Journal* on how the most effective communicators look directly into the eyes of their audiences.[55] This phenomenon is echoed in art. Most magazines will have their cover stars staring at you; most luxury adverts, especially with perfume campaigns, will do the same. Think J'adore de Dior without Charlize Theron looking you in the eyes.

This focuses on the eyeline of the star in relation to you. There are two types: demand and offer. When the star is looking directly at you as the viewer the purpose is to demand your attention. An offer is less direct where the star is looking outside the picture or at someone or something else in the visual rather than directly at you. This feels more contemplative to you as the viewer. This is also true on an interpersonal level – when someone is looking at us directly we experience a much more intimate or demanding feel to the communication than if that person is talking to us as we walk side by side.

Pay attention: a subject's gaze is usually deliberate – especially in advertising. The next time you're assessing your visual landscape, try to find an example of both demand and offer, and consider what each of their agendas is and consider how you feel in relation to them. Do you feel more compelled to take action with one approach over the other?

Up Close And Personal: Social Distance and Intimacy

Think of Edouard Manet's famous painting *Olympia*. The painting depicts a young prostitute lying naked on her bed looking straight ahead, with no floor and barely anything else in the room visible, just the bed, the woman and her servant girl filling the frame, which creates the illusion for the viewer of standing right inside the woman's bedroom. One feels very much like the suitor, who has just walked in and is met by her confrontational gaze; it is almost uncomfortably intimate.

Being literally up close and personal with the star of an image pulls you in, making you feel like you are part of whatever is going one in the picture, like the star is speaking to you directly. In adverts, this is often used to sell products to you, like jewellery or lingerie or even mundane things like yoghurt. Travel companies, for example, use a different technique. Usually they will depict people from a much greater distance, fully immersed in a fun activity or simply gazing at or walking through a beautiful landscape that will make you yearn to be there too, to reach the place where they already are, but you are not yet there.

Perspective

Horizontal Angle and Involvement

The easiest way to understand the horizontal perspective is to think of the sea. When you look out at the sea, it creates one straight line across your entire field of vision. Imagine this straight line running through the centre of an image to identify the horizontal relationships within an image.

The horizontal relationship between the viewer and the star of an image communicates a message. Where the star is positioned fully facing the viewer it implies that you are 'one of us', illiciting a closer connection with the subject. The alternative is an oblique angle where the star is off-centre or positioned at an angle to you as the viewer. This elicits more detachment in the relationship and implies that the star is 'one of them'. This distance between you and the star is happening on a horizontal axis.

As you go about your day try to spot visuals that use these two types of involvement angles. How effective are they in influencing you? Pay attention to that and decide if you want that influence in your life.

Vertical Angle and Power

To understand this point, remember images of the day the New York City Twin Towers fell on 11 September, 2001. We all know where we were that day. One of the most powerful and striking photographs from that day is an image of a man in sailing mid-fall as he leapt to his death from a fire- and smoke-filled Tower One of the World Trade Center. It was taken just after terrorists crashed two hijacked passenger planes into the towers.

The tower, still standing and vertical, and the man flying outward from it on the horizontal is a striking contrast between a horizontal and a vertical perspective – a (still-upright) steel building and an outwardly sailing human body. This pair of juxtaposed perspectives leads us to feel that the tower is so obviously huge and he's so obviously small that you come to relate to the fellow that's jumping from that tower – with one look, you register in your mind how big a disaster you are witnessing and how small and frail human life is in that context.

Vertical compositions that emphasise scale are used every day. From a business owner who wants to make his new storefront appear inviting by photographing it at eye-level, to the influencer who takes self portraits from a lower angle to make herself look taller, and the drone photographer who does the opposite and points his camera down at the ground to make people on the street appear small. These angles place you in a relationship with the subject to trigger an emotional response when you see the image.

We've all seen posters of fundraising campaigns for children's charities like UNICEF. Almost always the children on the posters look up at you, implicitly urging you to use your power to help them: they are framed to make the subject say, 'I am downtrodden and I look up to you to change that.' Alternatively, the image will be composed in a way that the depicted child will look straight at you, reminding you of your shared humanity, in a sense confronting you with the fact that it is only good fortune that makes the difference between your life and theirs. But never will you find a poster with this purpose depicting a child looking down at you. It is completely obvious that this kind of visual would make no sense in this context because the creator of the image is trying to establish a specific kind of relationship between you and the subject of the image that will convince you that you can help them. When you are looking up at a subject it tends to appear more powerful or imposing.

These perspective relationships also extend to what we see between different subjects in an image. The first relationship is always between the subject and the viewer but there's also a second dynamic between the star and other subjects within the image. To decode what those relationships are, you need only imagine yourself gazing from the perspective of the star – are they looking up at something or someone larger than them? Are they on the same level? Or are they looking down? How would you feel in their shoes and what does this tell you about the relationships within the image?

In short, when you see an image, these three angles are typically used to deliberately influence your response to it:

- High angle where the star is looking up at you, in this case the star is supposed to look as though they have less power than you do.
- Medium angle where the star is looking horizontally at you which is supposed to represent equal power.

- Low angle is where the star is looking down at you which is meant to show the star has more power.

As you go about your day, try to spot visuals that use these types of power angles. How effective are they in influencing you or impacting how you feel? Pay attention to that and decide if the message is warranted or welcome. You'll never look at a poster in the same way again!

Distance and Orientation

If I say 'pyramids' and you close your eyes, you will most likely imagine the visuals used in thousands of tourist brochures, the ones that present the pyramids as the tourist industry would like them presented – magnificent ancient structures, isolated in the middle of the desert. There is little doubt that the pyramids are a wonder of the world, but isolated they are not. They are just twelve miles from downtown Cairo. Cairenes Egyptians might be able to look out of their kitchen window while doing the dishes and see them!

Although we've covered some of the orientation information in composition in terms of how the star is oriented towards the viewer or other people or objects in the visual, there is also pre-verbal information being transmitted in a visual by the positioning of the elements in the visual.

The placement of the star in the visual allows it to take on different information roles. As in the example above, the angle used to depict a subject can hide as much as it reveals.

It is designed to convey a message about the subject. Consider where within the image the subject has been framed. Imagine the image broken into quadrants and consider the following: the star or information in a visual that appears on the left is considered 'given', familiar or common-sense knowledge, whereas what's contained on the right is 'new' or represents a problem or solution. In the Western world, whatever is positioned at the top of a visual is considered the 'ideal', it is often emotive imagery – what might be. Everything at the bottom of the visual represents 'real', factual, informative, down to earth. (This only works in Western cultures that read from left to right, top to bottom.) The middle is usually reserved for the nucleus of information and usually serves as a bridge between the top and the bottom.

Chapter 5.

Context Matters

In addition to the colours and compositional elements that make up an image, the context in which a visual is created or consumed shapes the visual message you receive. From the tense of a visual to the presence of symbolic objects and where it is located – all of these elements extend beyond the visual itself to shape the message you're receiving and, in turn, the visual narrative you build from these messages.

Tense and Time

Just like verbs in a sentence, visual messages have a tense embedded within them which contextualises the message you receive from them. Imagine King Charles having his portrait painted. King Charles is creating an image set in the present tense. He is using the image to say, 'Just so you know, I'm socially opulent. I'm King now. I have an abundance of wealth and status.' He's showing you that status through symbolic imagery within the painting. Images that were created as records of the current moment can thus be said to be in 'present tense'. Another common tense for images is the future tense. We see this a lot in advertising. An advertisement will present an image that you are meant to project yourself into. As an example, your house is crumbling around you and 'Here's a way to fix that.' Or 'Here's the place you could be living instead.' Or you might see an image of people enjoying a beer on the beach. These images show you a version of your future, one that can be made possible with their products. An example of a past-tense image is one that is created to represent the past – for example, Wes Anderson usually sets his films in the past, and he creates a visual sense of nostalgia using a careful attention to detail and colour grading. These visuals are past-tense – you are meant to interpret them as history even though they were created more recently than the setting of the image suggests.

The tense of the image matters a lot because it shows you what the objective of the image will be. Ask yourself whether the creator of the image was representing their past or their present, and if it is not a clear depiction of either one, ask yourself if the visual is trying to get you to know or do something right now; is it inviting you to see yourself in a later moment? A visual is always trying to communicate a message, so understanding the tense is key to decoding that message.

Questioning What You See

Crucially, when it comes to our visual consumption and narrative it is equally necessary to consider what's missing from the frame. In an age of fake news, when lies and conspiracy theories are spilling out into the real world, it would serve us all to consider what we might not be seeing and therefore the authenticity of what we are seeing. Fake news can mislead people and provide them with inaccurate information about events or situations, which can lead to incorrect decisions or actions, but we can also be misled by incomplete or falsified images. It can also cause confusion and uncertainty, making it difficult for people to understand what is really happening. Fake news, aided by misleading images, can also contribute to the polarisation of society by promoting extreme views and opinions. It can create a sense of division and conflict, making it harder for people to come together and work towards common goals; it can undermine democracy by distorting the truth and manipulating public opinion and even influence elections and political processes by making it harder for people to make accurately informed decisions.

For example, in 2018, a video was released that showed former President Barack Obama delivering a speech that he had never actually given. The video was a deepfake, created using machine-learning algorithms and audio from a real Obama speech. In the video, the fake Obama appears to be speaking fluently and naturally, and the audio is synced perfectly with his movements. It was created by a team at the University of Washington to demonstrate the potential dangers of deepfake technology, and how it can be used to spread false information or manipulate public opinion.[1] While the video was created for academic purposes, it quickly gained widespread attention and was widely shared on social media where it was decontextualised. It is an example of how deepfake technology can be used to create convincing and misleading content, and highlights the need for caution and scepticism when consuming online media in particular. The speed at which AI technology is evolving, I'm sure by the time you read this it will be easier than ever to create fake videos and images and use those to manipulate people.

Though the technology has changed, it is important to note that creating false or misleading images to influence politics is not a new phenomenon. A famous example from history is presented in a *New Yorker* article by Masha Gessen of 15 July, 2018, 'The Photo Book That Captured How the Soviet Regime Made the Truth Disappear' revealed how leaders who fell out of favour in the former USSR (Trotsky, Lenin, etc.) were erased from their history books and other physical records.[2] The USSR regularly erased individuals who had fallen out of favour from official photographs, deleting them from the historical record and making it seem like they never existed.

We have also seen image manipulation used to sway public opinion simply by deciding what is in the images being shared – no technological manipulation necessary. One of the first recorded instances of Western photo editing is the image taken by pioneering photojournalist Alexander Gardner in 1863, after the Battle of Gettysburg in America, a historical battle which marked the turning point of the Civil War. The image shows a young confederate soldier who has died in battle. But it's not real – it is staged. Sadly, the man *is* dead, that is real. And he did die in battle, but this is not where he fell. Gardner and his assistant moved the body about 40 yards and arranged the corpse, using his skill in composition to tilt the head towards the camera for greater emotional impact. He also put his own gun against the back wall, pointing up to add some additional gravitas to the shot. Gardner carried this with him as a prop to add to images for greater impact. With more than 50,000 estimated casualties, the three-day engagement was the bloodiest single battle of the conflict, and this image was designed to communicate the impact of the battle in a way that would engage viewers emotionally.[3] This next section will help you think critically about the images you're seeing to better resist manipulation.

Props

Gardner's gun was a powerful prop because it acted as a symbol for the war, indicating to the viewer why the young soldier died. The gun, paired with the young man's uniform, tells us that he was a soldier and that his death was due to political conflict. It made the cost of the war very real to the viewer of the image. Props can be used in visuals to emphasise a message or add drama or poignancy to the image. Sometimes they tell us something about the star or they link the image to an idea or well-known story. They can convey emotions through their presence even if the rest of the image isn't as detailed. Props are not necessarily items that have been added to the image for effect, like Gardner's gun, they might be objects that are actually there, but the image-maker's decision to focus on them turns them into a symbolic part of the visual message.

Case Study
Yalda Hakim

Yalda Hakim is an Australian broadcast journalist, news presenter and documentary film maker whom I've deeply admired for years and come to call a friend. She was born in 1983, in Kabul, Afghanistan, and currently lives in London, working as one of the chief presenters for BBC News. As a presenter Hakim is distributing world-shaping news on a day-to-day basis, informing audiences across the globe on the state of the world. While in Kabul, Hakim poignantly remembers looking at the barbed wire from her window in the compound, and finding it difficult to put into words the feeling of being there.

No matter how much she tries to help audiences understand what women in Afghanistan are currently going through, nothing better summarises the sense of being trapped than viewing that barbed wire. When staying in the local bureau office, and even though knowing that with her foreign passport she could more easily leave and get away, the visual of the barbed wire symbolised the nature of what was going on for so many. When she told me this over the phone, I remember I was pregnant and felt a deep pain in my stomach. Think about the symbolic message communicated by the barbed wire – it is sending a visual message with its presence, and choosing to make it the focus of an image changes your perception of that visual.

For a more deliberate example of a prop being used to send a message, we can turn to American news: in 2021, the world witnessed the United States of America's Former President, Donald Trump, and his use of the Bible as a prop as he stood outside St John's Church near the White House following the Black Lives Matter protests.[4] News reports had surfaced of him retreating to the White House bunker as the protest grew outside and he deliberately placed himself in front of the church holding an item with symbolic weight in order to send a message. In my eyes, and those of many readers and no doubt visual analysts, it was clear that the Bible was meant to speak directly to the religious and conservative electorate of Trump, symbolically aligning himself to them in that moment.

Source

The source of the visual will always tell us something about that image, especially when it comes to the visuals we consume online, either through websites or social media. Who creates or shares the visual can inform us a lot about the possible interpretation of the image. While visuals do have messages embedded directly in their colours, shapes, compositions and subjects, there's an additional layer of meaning that comes when you understand where the image came from. Knowing the source will help you understand how the image was constructed, and understand its intention and its objectives.

In the plate section of this book, you will find a photograph of a female being carried away by an unusually strong contingent of uniformed men. You may well be aware that the female looks like Swedish climate activist Greta Thunberg. But without knowing the source of the photo, you can't be sure that it is her and what the message is. Although the female (and it is Greta) is central in the shot, therefore the focal point, where is she and who are the men in uniform hauling her away? Away from what? It becomes a political statement when you source it at a coal mine, the site of a protest about the expansion of that German coal mine at the time. You will come to your own conclusions about its meaning when you view the photograph in isolation, but consider how your perspective changes if you believe it was taken by a climate activist or a representative of the coal mine? Consider what kinds of messages can be sent using this image. I believe there are different visual languages, and sources – the Thunberg photograph is key to speaking that language.

If you are an advertiser, you need to sell your product. So just knowing that you are looking at advertising can give you clarity on how the image is being used to try to manipulate you. You can assume that its objective is to convince you to buy. If you notice that an image is sending a controversial message, that visual is likely trying to sell you on a viewpoint. Identifying the source here can give you a clue about what they're trying to convince you of. For the viewer, it is still all about comprehending the source and the objective of that photo.

There are different languages, and knowing what they are allows you to approach visuals more consciously. If you identify the source, you also know what symbolism or what kind of language they'll try to use, because the ultimate objective is now something that you have identified for yourself. Not detecting the ultimate objective is obviously dangerous because then you don't know what the manipulation is.

In an article for CNN in February 2022, photographer Àsìkò warned that the Western world was the main source for most visuals we see of the African continent.[5] This drastically changes viewers' perception of the continent and as a result his art stands to counter the overtly Westernised visual narrative that we see daily. I believe that we need to actively challenge our visual biases by seeking out visuals from a variety of sources to ensure that we're seeing the full picture clearly. As you learn to identify where visual messages are coming from, and what patterns tend to arise within images from a similar source, your visual awareness of the messages you are consuming will grow stronger. Consuming images that question and counter each other helps you form your own opinion instead of unconsciously accepting either visual narrative as the whole truth.

Online Sources

There are 4.48 billion of us on social media, that's 4.48 billion voices. Imagine the sudden visual noise, it's overwhelming to say the least.[6] Previous generations – anyone born before Gen Xers – have never been exposed to so many visual voices all at once from a young age because the internet has only been around since 1983 and has been growing exponentially since then, changing all of our lives. This increased access to everyone's opinions and ideas could encourage both a sense of cohesion and a sense of divisiveness at the same time. While it is easier to find like-minded people, it is also easier to see ideas that you disagree with on a daily basis. That's what make us feel that we live in such a polarised world. At the same time, the way we use the internet means that all kinds of information on a wide range of subjects is being offered to us constantly, in no logical order. At the start of the day, when I open my social feed, it ricochets from my friends' individual successes to tragic international news: it's too much.

This competition for your attention has been linked to a shift in news media away from objectively reporting the news and focusing on truth and facts to a more entertainment-based approach that has a greater emphasis on opinion pieces and personalities sharing their take on the news. Money is made online via clicks and views, so the visuals and the stories that are the most sensational often make the most money regardless of whether or not the visual or story being presented is true or not.

AI will help unscrupulous sources generate more vis-uals and content to compete for your attention at a faster rate than ever and the current systems provide a clear finan-cial incentive for them to do so. With AI imagery out in the world, you have to go back to your questioning and second guess if the images you're seeing are real. Some countries have legislation around misinformation (for example, laws governing false advertising), but when we're online it's our responsibility to tell the difference between fiction and non-fiction, news and entertainment, clickbait imagery and real images corresponding to truth. There's nothing wrong with something that is fictional if you understand that it is not a representation of the truth – the trouble comes when images appear to be true but aren't. Image-creator Pablo Xavier encountered this first hand when he used Midjourney to create an AI image of Pope Francis wearing a puffer jacket – when the image was released online it became co-opted by third parties looking to criticise the Catholic Church.[7]

It is not always possible to know the source of a visual message but there are certainly more tools online that allow us to check them. Google Images or TinEye offer the capacity to reverse search an image. Going back to the image of Pope Francis in the puffer jacket – if you were to reverse search this image, you would find that Xavier had first posted it on a Facebook group called AI Art Universe and then on Reddit, a clear indication that it was articificially created. Reverse-image search is an example of where artificial intelligence tools are able to support us in finding a true source and reveal who has been sharing the image. Finding the original source is important because a common fake news strategy is to use a real image or visual but apply it to a different narrative. The original source is important, but so are the people who are resharing the visual because they may be doing so within a new context.

For instance, during the protests in Hong Kong in 2019, social media was flooded with images and videos of police brutality and violence against the protesters.[8] However, the Chinese government attempted to control the narrative by manipulating images and videos to present a different version of events. One example of this was the use of deepfake technology to create a video that appeared to show protesters attacking police officers.[9] The video was widely shared on social media and used by the Chinese government to justify their crackdown on the protests. It was later revealed that the video had been doctored and was not a true representation of events. In another case, the Chinese government released images of what they claimed were protesters holding weapons, including guns and grenades. However, Bloomberg revealed that those images had been taken out of context and were actually from a military reenactment event that had taken place years earlier.[10]

As another example, in 2020 Fox News used digitally altered images in its coverage of Seattle's Black Lives Matter protests in the Capitol Hill Autonomous Zone.[11] The aim of these protests was to reallocate 50% of the $409-million-dollar Seattle police budget into community programmes and services in historically Black communities to tackle the underlying causes of incarceration and police intervention before they start. Images shared by Fox News showed an armed man at the entrance of Capitol Hill, but the original photo showed an empty street. The images were manipulated to make the protest look more violent and chaotic than it actually was. The altered images were criticised by journalists and social media users for misrepresenting the situation. The incident highlights the potential dangers of image manipulation in media coverage.

However, sometimes the visuals that you're seeing are manipulating you on a softer and smaller scale. Professor Talia Stroud commissioned a study on polarisation and partisan selective exposure in 2020 in the *Journal of Communication*.[12] She demonstrated that people gravitate towards media outlets that match their political predispositions, including magazines and news organisations, and that this self-selection of sources that share existing political views can increase polarisation. For example, media outlets have political leanings and may selectively report on certain topics or events to support their agendas (e.g., left-leaning news outlets may emphasise stories related to social justice issues while right-leaning outlets may focus on crime and immigration), leading to people who consume those stories becoming more convinced that the biased perspective they are consuming is right. This means that when the visuals you are consuming are overwhelmingly coming from sources that share the same biases – which you might naturally gravitate to because you agree with them – you risk being more deeply entrenched within that biased worldview and miss stories that are not being priorised by those outlets.

When it comes to what you see on your social media feeds, you might be receiving misleading visuals simply because people are not taking time to think critically. Psychologists David G. Rand and Gordon Pennycook found that there's a disconnect between what people believe and what they share on social media – the sharing of fake or misleading information is largely driven by inattention, lack of careful reasoning and lack of relevant knowledge rather than partisanship or intentional misinformation.[13] Taken together, these two studies show us that we must continue to question the visual messages we're receiving even when we trust the source because even trustworthy sources may only be telling us part of the story and the people around us might not always realise that an image is misleading at first glance.

When it comes to thinking about how the source of an

image changes its meaning, it isn't always enough to just think about the individual or organisation that has made the visual available – the platform or medium you're accessing the visual through shapes the visual message by creating the context in which you encounter it. If I pick up a magazine, I know that a lot of the visuals I see are advertisements because selling that ad space helps fund the production of the magazine. Many social media platforms, even if it isn't obvious, are designed to keep you scrolling so algorithms will suggest visual content that will keep your attention fixed on the platform. The platform is acting as a secondary source with its own agenda: to manipulate you into looking at the next visual, and the next, and the next. While the image might not necessarily have been created with the purpose of being liked and keeping you on social media, the platform frames the image as something to be liked, engaged with, consumed and then passed over in favour of the next visual.

Algorithms are used by platforms such as Facebook, Twitter, YouTube, Instagram, TikTok and many other online and mobile-based social platforms to show you the content you're most likely to keep looking at. They can create filter bubbles, which is what happens when users are exposed primarily to content that aligns with their existing beliefs and interests. The technology on these social platforms also knows when you slow down your scrolling, and learns your preferences and interests when you comment, share or like. This data fine tunes the algorithms so you see more of what catches your attention[14] and this can sometimes have tragic consequences.

In October 2022, a British judge decided that Meta, the

owners of Facebook, Instagram and WhatsApp, were responsible for contributing to the death of seventeen-year-old Molly Russell.[15] Molly had been suffering from depression, and the Instagram algorithms picked up on this and fed her more content that made it worse. One psychologist who spoke in court said when reviewing the imagery that Molly was shown, it caused her to lose sleep for weeks.

In 2021, thousands of pro-Trump extremists stormed the Capitol Building in the US to try and prevent the confirmation of President Elect Biden.[16] Many in the crowd were there because they believed that the election had been stolen, even though more than sixty court cases found no evidence of wide-spread voter fraud.[17] Social media was awash with groups that were involved in peddling this voter-fraud conspiracy theory, making it seem to them like credible news. If these people were predominantly getting their information from fabricated images and videos then it's not too surprising that they were convinced. We are all susceptible to this type of manipulation if we limit what we see and pay attention to. This is especially true online where anyone can say anything. There is no law to prevent the spread of lies and no consequences for those who do.

It may be time for us to be more vigilant around the intentions or potential intentions of the visuals we consume. If we have had a strong reaction to something, this is almost certainly deliberate, so before you click on 'share', it is worth double checking the source to better establish the likely intention of that visual. Periodically pause to take stock of the patterns in what you see – how are the algorithmns you interact with trying to keep your attention? Pay attention to the kinds of emotions that this visual landscape is triggering for you and decide if this is the kind of emotion you want to experience every time you go online.

Exercise
Burst Your Bubble

To get outside the filter bubbles that algorithms have created around you, there are a few things you can do to bring different kinds of visuals into your feed. When it comes to search results, you can use VPN and clear your internet browser cookies to get less targeted search results. I recommend doing this regularly as simple internet hygiene. The best way to improve the algorithims is to actively seek out new visuals. If you look at content targeted to Millennials, google some retiree content; if you never game (that's me – I've never played a video game in my life), search for video games. If you feel YouTube is recommending you only one type of content, search for topics that you literally know nothing about and watch a few of the videos that come up. Think of things that are just very different to you or that 'belong' to a different age demographic from you. Think of the type of voices you're not usually hearing from and seek them out. So if you see too much similarity, try and break that similarity by literally going in the opposite direction and complicating what the algorithmn should show you. That's the best way to break any form of filter sending you to limited bubbles of content. You can 'break' the algorithm.

Picture Perfect

Not only do the visuals we consume on a daily basis shape how we see the outside world, they also shape how we see ourselves. Nicholas Mirzoeff coined the word 'visuality' to define this phenomenon. How we define our existence to ourselves is shaped by what we visualise and the images we have access to, which are in turn shaped by authority and are influenced by who is in power. Bourgeois and working-class people only started to appear in painting in the 19th century, but thankfully we've come a long way since then. Today the power to create visual constructions that include all of us has grown exponentially and we have access to more representation than ever before, but we still live in a world where our visual consumption can limit and affect how we see ourselves. It's no secret that many of the images we see are of supernaturally beautiful people. Photoshopped images can perpetuate harmful beauty standards and reinforce the idea that a certain body type or appearance is necessary to be considered attractive or desirable and these images sink into our subconscious and get under our skin. This can have a negative impact on self-esteem and can contribute to feelings of inadequacy and shame. Excessive exposure to unattainable beauty standards has been proven to lead to a range of mental health issues, including anxiety, depression and eating disorders.[18] With hyperrealistic filters that can alter our appearance on live video available at the touch of a button on our phones, it is increasingly difficult to spot edited images and videos. While we can detect some signs of editing (e.g. inconsistencies, blurriness or pixelation when enlarged, repeated patterns, etc.), depending on the level of skill with which it was done the only thing we can really do is remind ourselves and our children not to take everything we see in print and online/in social media at face value and most importantly to stop comparing ourselves to these images of apparent perfection.

The beauty standards that we hold ourselves to differ between cultures, but for all of us who spend time online, any conversation about beauty standards has to consider the influence of pornography. In 2023, the global porn industry was worth around $97 billion.[19] Bigger than Netflix. Porn has had a huge influence on body image because, while there is admittedly a porn genre for absolutely every body type and preference out there, mainstream porn is most certainly responsible for various expectations of what our bodies should look like. Big boobs are obviously the first thing that comes to mind, quickly followed by bodies being entirely purged of body hair. We all want to be attractive to other people, but porn overwhelmingly prioritises a singular ideal that goes beyond what individuals find beautiful, making people question whether they can even be sexy if they don't fit the image. Porn has also been linked with a rise in obscure procedures like labiaplasty and anal bleaching, with individuals (including very young girls who might view this type of material) thinking there is something wrong with them if their bodies don't look the way they see them in porn and on social media. One of the first uses of deepfake technology was for creating pornographic films that appeared to include celebrities before it bled into other areas of content production.[20] To say nothing of how horrible it is to use someone's likeness against their wishes in this objectifying way, it also points to the other place in our society where a lot of our images of beauty come from: celebrities.

A lot of our ideas about what is beautiful in our culture can be identified by looking at who we make famous for their beauty. Beauty trends such as getting lip filler can be linked directly to how we're influenced by seeing specific people as beauty icons. In fact, according to one investigation by the reporter Monica Corcoran Harel for *Elle* magazine, 'almost every expert [she] asked about the filler era name-checked the Kardashian-Jenners'.[21] A paper by Kiamani Wilson published in 2018[22] examined a sample of United States plastic surgery data from 1997 to 2016 from the American Society of Plastic Surgeons, supplemented with data from Google trends and from the Federal Reserve Economic Database. It revealed that post-2013, after Kim Kardashian and Nicki Minaj rose to prominence (including Kim Kardashian's famous 2014 'Break the Internet' cover of *Paper* magazine, where she displayed her rather voluptuous buttocks, well-oiled, sleek, and far too perfect for real life), there was a 51% increase in breast augmentations, buttock augmentations, buttock lifts and tummy tucks compared to other cosmetic surgeries. What the Kardashians and Jenners, and people of visual influence like them, portray to the world visually impacts the minds and actions of a large portion of the planet. At the time of this book's publication, Kylie Jenner alone has reached over 400 million followers on Instagram and collectively the family have amassed over 1.2 billion followers.

The traits considered beautiful by porn and celebrity culture get replicated and multiplied by social media so that they influence us and affect our self-image even if we aren't consuming pornography or tabloids firsthand. In the monograph, 'Effects of social media use on desire for cosmetic surgery among young women', published in 2019, the authors demonstrate that social media impacts body image and self-confidence, especially in young women, which is thought to be at least part of the reason behind a 70% rise in plastic surgery in young women, even girls in their teens in the Western world.[23] This study examined the impact of viewing images of attractive and thin women on Facebook on young women's body image and mood. The study recruited 112 female undergraduate students from a university in Australia. Participants were randomly assigned to view Facebook profiles of either attractive and thin peers or neutral photos of friends. Results showed that those who viewed the attractive and thin profiles experienced more negative body image and mood compared to those who viewed the neutral profiles. This endless search for beauty has also seen a dramatic rise in the use of Botox and lip filler in young women as, according to the American Society of Plastic Surgeons, it became the most popular cosmetic procedure in the world in 2020.[24] This is thought to have been triggered or certainly fuelled by Kylie Jenner's admission that she had lip filler at sixteen years old.

Thankfully, it is not all bad news for us – we have the

power to change how visuals impact our self-image by changing our visual landscapes. A study published in 2020 by psychologist and researcher Marika Tiggemann found that the increasing trend for body-positive content on social media can have beneficial influence on users. Participants were 384 women aged between eighteen and thirty, randomly assigned to view Instagram images of thin or average-sized women containing either body-positive captions or no captions.[25] In contrast to prediction, the body-positive captions had no effect on body dissatisfaction or body appreciation. What did have a significant impact was the type of image. Tiggemann found that average images resulted in less body dissatisfaction and greater body appreciation than the thin images. She also found a significant three-way correlation that for women who thought of themselves as thin, body-positive captions on average images led to greater body appreciation, but lower body appreciation when attached to thin images. The results suggest that the visual imagery of an Instagram post is a more potent contributor to body image than any accompanying text. Presenting a more diverse array of women's bodies on social media is likely a more effective way to foster body satisfaction and appreciation. While a lot of research focuses on the effect of social media on young women, a beautiful example of body positive content targeted at men is the Instagram account ArrestedMovement belonging to photographer Anthony Manieri. He produced a series of nude portraits of men of all body types, celebrating the beauty of diversity. The response he gets for these pictures is one of overwhelming thankfulness, with people feeling seen, represented and understood and able to appreciate and love their own bodies, in some cases for the first time in their lives.

This shows that regardless of our gender we can counteract the harmful effects of only seeing one kind of body by making sure we're exposed to positive representations of a wide range of body types. In addition to diversifying your feeds, I recommend that you find real-life spaces where you see a wide range of bodies in a non-sexual context. For me, I find the best place for this are the swimming ponds you find dotted around in the parks in London. Unlike the beach, the swimming pond doesn't feel like a place where people are trying to be sexy, they're all just there to swim in the cold water. Everyone seems to give a sigh of relief as they look around at other 'normal' people that says, 'We're okay, we're right in the mainstream, we really are in the same boat aesthetically as everyone else!' It's the total opposite of what social media, advertising and publicity will give you as they unite to inundate you with an unachievable ideal of beauty.

Location

Location is everything for street artists. Looking at their work can help us understand how where we see something shapes how we understand its visual message. My first boss, Steve Lazarides, specialised in supporting street artists – he discovered Banksy and JR. These artists specifically choose to place their work in places you wouldn't expect to find it as a way to challenge you on the visual world you see every day. Perhaps you remember one of Banksy's most famous artworks, *Flower Power*, which features a protestor caught in the act of hurling a bouquet of colourful flowers, not the Molotov cocktail you might expect. He painted it on Jerusalem's West Bank barrier to highlight his support towards Palestinian rights. Remove the location here and you have lost most of the meaning of this visual. Location shapes the visual messages we consume every day, and shapes our visual narratives.

When we are assessing the visuals we consume, location is important on two levels. First, there is the location of the visual, the backdrop used. Why that location? What is the creator of the visual seeking to tell us through that choice? Again, is it real or even true? The second aspect of location is your physical location as you see the visual. As we will hear from Adam Nathaniel Furman, the space you are in when you encounter a visual gives it meaning. Seeing inclusive artwork in a public space can make you feel more welcome in that space. This principle is true for all kinds of visuals, not just art.

Case Study
Adam Nathaliel Furman

Visuals for artist Adam Nathaniel Furman have always acted as a tool to communicate something when words couldn't. Hailing from leafy Hampstead, North London, and a family of Jewish immigrants, their childhood home was filled with a number of curiosities. Fantastic works of art and furniture lined the walls and floors, made from fine materials such as marble, ivory and carved wood. Yet finery was matched with the common and everyday, like doilies – with the home appearing 'kitsch', as the artist describes it. Yet 'kitsch', like 'clutter', did not possess a negative connotation in the vocabulary of their household.

Instead, Adam was enchanted by the appearance of all of these different objects; enthralled by their varying appearances, meanings and provenances. This was something they further nurtured with their aunt, who would take the artist on 'antique-hunting' expeditions during the weekends, where they would observe a vast assortment of objects. Adam recalls their aunt's house as being 'rammed with mummified cats, strange stone objects and novels', with all the items creating a 'story atmosphere'.

From such exposure to the eccentric and eclectic, the artist treated their childhood bedroom as a space for total 'creative freedom'. A subscriber to *National Geographic*, they would dissect the magazine, glueing its pages to the walls and ceiling of their bedroom, gazing up at maps from around the world. In later years, this act of collaging turned into painting, where Adam would paint murals onto the walls, depicting abstract patterns and portraits of people. The curation of objects in the room was also ever-changing, with the artist wanting their visual environment to 'constantly appear different'. Creating art, as well as looking at objects in the home, was a form of escapism for the artist in their youth. Bullied for being queer when they 'dressed differently [at school they] would get abused', so instead they used their portfolio as a creative outlet to explore their queerness, due to it providing a certain level of 'privacy'. As well as this, they were troubled with learning difficulties, so it was colour which first spoke to them – 'colour was a prediscovery of language. It wasn't a social or decorative thing, it was a mode of expression'. In this sense, Adam truly created a visual environment in their youth to escape the torments of daily life; they created art to design a visual world they could see themselves in.

Moving on from their childhood home, it was Soho, the mecca of the LGBTQIA+ community in London, which was a great inspiration for the artist. Adam was particularly enamoured with the entrance of Soho-Tottenham Court Road Station – which they profess is referenced 'in every project [they] do'. Their greatest memories there were witnessing the iconic Eduardo Paolozzi's mosaics (completed in 1985), which bedazzled the underground; famously a source of inspiration for many other queer artists in London, such as the artist Francis Bacon, who frequently visited them. Adam recalls watching drag queens ascending the flights of stairs at Tottenham Court Road like angels, where they would stop to get their photographs taken. This site for the artist was a 'dingy, sexy and glittery entrance to Soho queer life'.

The London Underground decorative schemes, 'with their enamel panels, posters and ceramic graphics', like the many objects of their childhood, were of great importance for the artist's visual consumption, noting the underground was a site embedded with hyper 'interesting decorative arts, which greatly inspired me'. Yet in 2015, as part of a larger effort to build the Elizabeth Line to renovate the London Underground, Paolozzi's mosaics were stripped, with the majority of them being shipped off to Edinburgh University. The renovation was of great stress to the artist, with them exclaiming, 'There is no single visual I see which is negative, it is only the absence of visuals in spaces [like the underground] which I see as negative.' The artist feels frustrated by the new design of the CrossRail line, due to it being devoid of architectural ornamentation, posters or ceramic graphics – instead conforming to a cleaner, 'more clinical' aesthetic. The artist believes that a lack of visuals results in a complete absence of 'visual stimulation'.

Despite the growing trends of minimalism in London, Adam is still working towards crafting a visual environment filled with colour and curiosity. They have most recently been commissioned to create a 57-feet mosaic (the largest-ever mural in London), in collaboration with the London School of Mosaic, alongside London Bridge Station. In this sense, just like in their youth, Adam continues to use colour, and art more broadly, as a tool to craft a visual environment which

> speaks to them. They are introducing a queer narrative
> to London's visual environment in a subtle but strong
> way, allowing those who feel they don't have a space to
> feel seen – in bright, bold, brilliant colours.

Nowadays, we are targeted by advertising everywhere we go, from the comfort of our homes with our phones and screens to outdoor advertising. A great term I've come across to describe this is 'visual capitalism' – which refers to how, particularly in dense commercial areas like Times Square or downtown Hong Kong, the majority of the visuals that we see were created with commercial goals and uphold a capitalist worldview. When it comes to commercial messages, marketers can play with different locations, both online and in the physical world to get to us.

One of the best insights into online marketing I've received happened back in 2019 when I was gifted mentoring sessions with a retail company. The company training me explained how they track a 'customer journey' through their site, and showed me the mechanism that ensures that every time you open your computer or your social media apps, if you have clicked or spent time looking at a specific clothing item, it will be re-advertised to you elsewhere online. Our current world is a dream for advertisers, politicians and marketers because online marketing makes it easy to place visuals targeting you in different locations. Unaware of this, you end up consuming them repeatedly and, potentially, following the course of action that they would hope for you to take, whether that's consuming an item or an experience or voting for a political party. Where you see the visual matters because it will impact you in different ways and repeatedly seeing the same message slowly makes it more convincing – this is true of both virtual and physical spaces.

Social media platforms allow advertisers to create highly targeted campaigns based on user data registering their interests from their visuals and videos on social media. Here, what we look at becomes data used to feed us more targeted visuals and seek our attention wherever we spend the most time. Print media advertising, such as newspaper and magazine ads, and TV advertising are still powerful tools for getting our attention, but they don't give advertisers as much data about who saw the visual, how long they viewed it, and even where they were when they saw it, but they are very good for reaching us during our downtime at home, surrounded by our families, when we're at our most relaxed. Likewise, outdoor advertising is less targeted, but it has the advantage of being highly visible and can create a strong visual impact, drawing in a wider range of viewers; however, the effectiveness of outdoor advertising is limited by the location, the size and design of the advert, and the length of time it is displayed.

Key locations for ad space are as varied as the advertisements themselves. While airports are good to connect with senior-level executives, affluent consumers or global opinion leaders, areas like Times Square in New York, Piccadilly Circus in London or the Shibuya Crossing in Tokyo work best for a varied, mass-market audience, since not only do thousands of pedestrians walk past these billboards every day, but these specific locations have also become tourist attractions not despite, but because of their billboards. Another premium location is in San Francisco, positioned just as you come off the Bay Bridge, because it's the main road into town. It carries some 280,000 passengers daily, and in 2020 motorists lost an estimated 600,000 hours per mile due to congestion, which gave them ample time to look at the visuals on the billboards.[26] The ubiquity of these advertisements has an impact on our wellbeing because they push images into our visual landscape without our consent, altering our visual narratives for us.

In light of how intrusive advertising can be, governments are starting to regulate our exposure to the advertising we see on our streets and impose more public art projects. Countries like Switzerland and Germany have fairly strict regulations for advertising signs. They have rules that prevent ads from obstructing the view onto parks and landmarks, and in residential areas advertisements are only allowed if directly in the vicinity of the shop that they advertise for. In these countries there's a strong sense that the aesthetic appearance and visual cohesion of the cityscape is just as important as the right (and necessity) for business to advertise. As of 2023, four US states, Alaska, Maine, Hawaii and Vermont, have also implemented measures to curb pulblic advertising.[27] These states currently have a complete ban on billboards in order to not disturb the natural beauty of the landscape. Rhode Island and Oregon have prohibited the construction of new billboards. In 2020, Kyoto, in Japan, passed laws to enforce restrictions on how businesses can visually display themselves and advertise outside. If you have an old monument that you're restoring in the UK, you can paste the big fat logo of a company all across the billboard that is basically protecting the restoration monument. Not so in Italy and France, where the monument is an image and the logo of the company can only be at the bottom or in a small size. All of these measures are being taken to help protect your physical visual landscape from visual capitalism.

Another type of advertising that is being legislated is the number of advertisements you'll see on TV. In the UK, there needs to be a time-lapse in-between ads of at least seven minutes. In the US, they allow you to advertise every two minutes. Some countries, including the UK, US, Australia and the EU, also have legislation requiring social media channels to disclose when a post is a sponsored advertisement to help make it easier for users to identify ads and limit their consumption.

Exercise
Create an Ad-Free Space

Think about where you're seeing different kinds of visuals during your day. Are there any places where you don't see any commercial visuals? Try creating a physical space in your life that you go to every day where you don't consume any commercial visuals for one week. This might mean banishing your phone from your bedroom! How do you feel before and after this experiment?

Online Visual Consumption and Instant Gratification

While there is no shortage of uplifting and wonderful content available online and through social media – many art institutions from the MET in NYC to the Rijksmuseum in Amsterdam have vast digital collections available instantly online to anyone in the world – these visuals compete for our attention with everything else the internet also has to offer. Our online visual consumption, especially through social media, is shaped by the immediacy of endless new visuals and the constant stimulation that comes from interacting with the internet. Receiving likes on social media activates the same reward centres in the brain as receiving money or eating chocolate[28] and social media use is associated with increased activity in the brain's reward system and decreased activity in the prefrontal cortex, a region associated with decision-making and impulse control.[29] This can cause us to compulsively project and consume visuals that easily and immediately gives us a rush of dopamine. It seems bizarre and also incredibly sad that celebrities who post endless selfies have millions more followers than one of the greatest visual institutions in the world. Kylie Jenner (as one example of this) has 412 million followers on Instagram, a substantial number compared to the 5 million followers of the Louvre Museum in Paris, which can be visited in person or digitally through the visuals they display online.[30] The Louvre is a treasure trove of visual beauty and endless uplifting or contemplative content; what would you say about Kylie's content in comparison or contrast?

Back in 2019, I decided to test if the higher demand for narcissistic visual content on social media platforms was real. I wanted to test how posting self-centred visual content would affect my Instagram engagement, so I posted a scantily clad picture of myself. It was the same kind of image that is commonly seen on many Instagram models' pages: I was standing on the edge of a swimming pool with the blue of the water meeting the horizon. Wearing only aqua-coloured bikini bottoms, my arms were raised up as if I were reciting an incantation or doing a sun dance. Standing on the tips of my toes, my calves looked leaner than usual and my bottom perkier.

At the time, I had 24,000 followers and my average post often received 100 likes. Within just a few hours of posting, this became my most viewed picture by 76%. By this point I had been posting on the app for six years, so that is a huge jump in engagement given all of the photos I had posted before this one, including my post about having recently been named in the Forbes 30 Under 30 list of contributors to Art and Culture, which was a huge honour for me. And yet this picture of my bottom was more popular than any other achievements or meaningful moments that I had ever posted at the time.

This social experiment didn't only confirm my hunch about what kind of content now performs well, its impact was actually two-fold: it also gave me greater empathy for the people who are posting this type of content (more on that later). The takeaways from my social experiment aren't merely anecdotal or subjective. A wider study has also demonstrated what I found myself. A research team led by Judith Duportail in partnership with the European Data Journalism Network found that 'Instagram prioritises photos of scantily clad men and women, shaping the behaviour of content creators and the world view of 140 million Europeans in what remains a blind spot of EU regulations.'[31] They went on to add, 'Posts that contained pictures of women in undergarments or bikinis were 54 per cent more likely to appear in the newsfeed of our volunteers. Posts containing pictures of bare-chested men were 28 per cent more likely to be shown. By contrast, posts showing pictures of food or landscapes were about 60 per cent less likely to be shown in the newsfeed.' This is especially toxic for women, where the visual consumption seems to encourage and reward the female body but not the female brain. Add in the online abuse and misogyny that female leaders are subjected to and it's little wonder that girls don't want to be leaders. The year 2022 saw an increase in cyberbullying against female leaders which correlated with a decrease in the ambition of young girls: 21% of young girls when interviewed by the *Guardian* in November 2022 said that they would prefer a less ambitious job to avoid a potential backlash or cyberbullying – indicating that this phenomenon can have a negative impact on people who see themselves represented in objectifying and cruel visuals.

Cyberbullying isn't exclusive to young girls: in 2023 the Australian eSafety commissioner reported that nearly one in five children received some form of direct cyberbullying during lockdown, primarily via TikTok, Instagram and

SnapChat. Online abuse using targeted visuals wasn't just limited to young children or young girls, as 90% of the 1,700 reports of image-based cyberbullying they had received in the first quarter of the year were from young men aged eighteen to twenty-four.[32] While we don't all receive image-based abuse, we are all impacted by the harmful messages that make it into our visual diet.

The emotional rewards of a self-centred social media persona are seemingly instant, but as an adult with very specific professional aims, I don't feel swayed by such a prospect. But what if I was posting for personal purposes? Or if I was a teenager still trying to work out who I am and what is important to me? What if, like so many of us, regardless of age, I wanted to feel that rush of engagement and appreciation through likes? Most of us want to be liked, it's human nature, but what if the content that facilitates engagement and the appreciation we crave doesn't make us feel good? Or invites unwanted criticism or cruelty from others? Often, the content we are compelled to share simply amplifies our anxiety and reduces us to an empty stereotype. Even if we get the likes, we don't necessarily feel that good about them.

Everyone is talking about TikTok these days because it has succeeded in capturing our attention even more than Instagram and Twitter. According to *Time* magazine in 2022,[33] teenagers in Europe report spending up to six hours a day on the app. When a user opens the TikTok app, the algorithm shows them a video that it believes will be of interest to them based on their previous interactions on the app, including videos they've liked, shared, or commented on. The algorithm also takes into account the content of the video itself, including the captions, sounds and hashtags used. It analyses the video's performance in terms of likes, comments and shares, and then it uses this data to determine how to promote the video to other users. Additionally, the TikTok algorithm takes into account the user's device settings, such as location, language and device type. This allows the algorithm to show videos that are relevant to the user's interests and location. Very few people read the captions on Tik Tok which makes it a fully visual social media app, feeding us twelve times more visual content per second than Instagram does.

This increase in our visual intake – this is not just about TikTok users, but refers to all of us – is the reason why I'm so determined to bring an awareness of our visual consumption to as many people as possible. For many of us now, when we watch TV, we also have our phones in hand. Our number of devices has multiplied, which would only imply that the number of visuals, specifically online or internet-generated visuals, would have multiplied as well and are vying for our attention. You've now got your computer (laptop or tablet or desktop), your smartphone and your TV right in front of you – and perhaps a second smartphone issued by your employer. You're just multiplying constantly the number of devices, which also comes with additional visual intake.

Exercise
Completing Your Visual Audit

The first part of this book has been about increasing your personal awareness of your visual consumption and performing an audit of your visual environment to understand what kinds of visual messages you are consuming throughout your everyday life. The next part will give you some ideas about changing your visual environment with a view to improving your wellbeing in yourself, in your home or office, and out in the world.

But first, take a moment to go through all of the questions you've answered along your journey and use the following set of questions to create a full picture for yourself about your current visual narrative and where you'd like to change your visual consumption:

1. Can you identify any patterns in the kinds of visuals you see regularly?

2. When and where do you tend to consume visuals most?

3. What kind of visuals are having the biggest positive impact on your daily life?

4. What kinds of visuals are having the biggest negative impact on your daily life?

5. Are there specific spaces that make you feel good? Can you identify any visual traits they have in common?

6. Are you able to identify where most of the visuals you see are coming from?

7. Are you receiving visual messages that come from a diverse group of sources?

8. Do the visual messages you consume paint a full picture of the world? What perspectives are missing?

9. Do you feel like your visual consumption reflects who you are? Do you feel like it reflects who you want to be?

10. Are there any patterns within your visual consumption that you want to change?

11. When you describe yourself and your life, what words do you tend to use? Is your visual consumption pushing you to compare yourself negatively to others? Do you often see visuals that reflect lives similar to your own or positive representations of aspects of yourself that you're proud of?

Part II

Your Visual Detox

Chapter 6.
Taking Action

Having done your visual audit and learned more about how the visual messages you consume impact your life, it's time for the visual detox. This is where you can change your visual narrative when it no longer serves you or fails to create wellbeing, a feeling of harmony, or comfort for you. The next chapters will explore what you can do in your own life, at home, online and in your community to create a more nourishing visual consumption which can improve your life.

Starting with your personal spaces, you will learn ways to transform your visual narrative with colour and light, nature, textiles and walls, and how clearing clutter can be transformative. We continue with tools for visually detoxing your online or digital life. Many of us who are screen-bound for our work don't believe we can take control of those visuals. You will examine with me what can be done with unwanted ads or content and how to manage needed content and visuals.

Building simple habits into your routine to bring more positive visual elements into your daily landscape can significantly improve your life and support your mental health. I have a daily visual routine that I have practised religiously since I was seventeen years old and left home for the first time. Up until I studied the positive impact of visuals for mental health as part of my job, I saw it as a mere quirky routine, but now I see how this has helped my mental health for years. At night, I always sleep with my curtains open so that I can wake up to a view and the natural light in the morning. Daily, I tidy my house the night before so that I can wake up in a visual environment that I enjoy. Then, as I start my day, I always make myself a cup of tea whilst sat in my favourite room of the house (even now with two kids I still do it, although I can't promise that I have time to drink it entirely) and surrounded by the artworks I love. Even when it rained, I would always prefer to walk or cycle to where I needed to go so that I could shape the visual environment I saw on my way to work.

Nowadays, I cycle past the streets behind the British Museum to bring my eldest to school and it fills me with joy. We stop at our cute bakery in Fitzrovia and he chooses the pastry that he finds the most visually appealing and then we pick our favouite buildings on the way. If we find a leaf we like, we stop and we collect it. I love seeing the parks early morning, the Georgian architecture and the little retail shops. I would always prefer a longer route that makes me visually happy over a shorter one, and I would always put on clothes that bring a smile to my face, even if it's rarely practical. Deciding between practical versus what is visually inspiring is a true debate in my home and visuals have most definitely won so far. Adding these actions one by one I feel like I am creating a visual environment that makes me happy daily. It also has a positive impact on my family: my mother-in-law calls our home a magical place and my sons are used to walking an extra mile to be visually happy.

I constantly read that we need to optimise everything,

run while listening to a podcast (which I can never manage to do!), exercise while we clean our houses and practise all kinds of diets, but we rarely include the visual aspect of our lives in this quest for optimisation. Understanding that you can have a voice and a level of control in your visual environment feels very empowering (and a lot simpler than cleaning while exercising).

Chapter 7.

Transforming Your Digital Landscape

We now live in a world where 76% of the population in advancing economies owns a smartphone. Remarkably, 45% of the population in emerging economies also own a smartphone.[1] This gives billions of people, from all socio-economic backgrounds, a huge opportunity to curate their online experience – including, most importantly, their social media feed.

Technology has, in many tangible ways, corrupted our visual consumption but that is largely because so many of us don't yet appreciate the impact our visual consumption is having on us. The fact that you are reading this book means you are already aware that what you look at all day influences your mental and physical health.

The good news is that the same technology that is currently so easily corrupting our visual consumption can also be tweaked – allowing us to reclaim control over our visual consumption and alter our visual narrative.

Your Detox Plan for Your Digital Time

Many of us need to spend a lot of time in front of our screens for work. Yet hours of our day are spent in 'leisure screen time' too. Since we can't fully control what we look at during our work hours, these two categories of time need to be addressed separately:

Digital Detoxing at Work

For many of us, a significant part of our work must be done every day on our computers (and for some, on the smartphone and other devices). But do you really need to check your business emails every ten minutes? Many executives have succeeded in reducing their screen time by telling their teams and clients they will only check emails twice daily – depending on your job, you might be able to block off parts of your day for dedicated offline work.

Other strategies such as 'digital tasking' do something similar by focusing on grouping tasks to certain time segments every day, helping you to avoid interruptions and to work more efficiently. For example, can you do the new data entry every day at 9.15 and 2.15? This might also have an added benefit of streamlining workflows for the people who feed you that data: they will know when you are putting together the report and can plan when they submit to you on that basis.

Exercise
Work-Based Digital Audit

Do an audit of work-related computer tasks with the goal of having your eyes on the screen for less time every day:

- Are there any tasks that you currently do online that can be done offline?

- Are there visuals such as pop-up email notifications that are causing you to be more stressed or more easily distracted?

- Are there certain times of day when you can implement regular breaks from your screen and walk to a 'visual break'?

- What do you look at during your breaks? Do these visuals help improve your day?

Digital Detoxing at Home

While a work-time visual detox might be subject to what your job is and who you work for or with, you have more control over your leisure digital time, but this can be easier said than done because of how good online advertising and social media are at capturing your attention.

First, let's start with online advertising:

Imagine that you open up your desktop and head on over to the *Guardian* to read the latest news in the arts and culture section. What should feel like a moment of sanctuary, or bliss, becomes corrupted by a series of pop-up ads, with garish, visceral images of people losing '2 stone in 2 weeks!', or meal-replacement shakes, surely containing ingredients no human should be digesting.

What had originally felt like a pilgrimage into fantasy has now become obstructed by a series of images no one wants to see. These advertisements, colloquially known in the cyberworld as 'annoying ads', can have a detrimental effect on our visual consumption. They are non-consensual, and as such, can distract us from what we were originally searching for, or worse, can impede our own mental health by having to witness imagery which may play on our insecurities – such as weight loss.

Yet with the arrival of the advertisement blocker, people are finally able to take back control of what visuals they're looking at online. The advertisement blocker first arrived on the scene with the introduction of AdBlock, a browser extension developed in 2009 by Michael Gunlach. Designed for Google Chrome, Apple Safari, Firefox, Opera and Microsoft Edge, the browser blocks various advertisements and is currently the most popular advertisement blocker today. As of late 2019, it has been recorded that over 700 million people are currently using advertisement blocking software, with over 81% of Americans using the softwares.[2] The desire of users to have ad blocking software primarily stems from many online advertisements presenting negative content, as confirmed by a 2020 AudienceProject survey, which revealed 47% of Americans felt negative towards online adverts, with only 10% feeling online adverts had a positive effect on them.[3]

For the average user of AdBlock, the browser blocks over 80,000 banners, adverts and pop-ups annually, with that averaging out to 200 blocked adverts a day.[4] Yet, beyond the average, some users of the browser have been blocked by more than one billion adverts in the past few years. This is undoubtedly an astonishing number, and confirms how advertisement blockers are changing our online visual environment completely. What could go from being a moment of vex, angst or melancholy to seeing a series of advertisements which do not apply to you, is now cleared through the use of the blocker.

Social Media Feed Curation

Now that you've removed the bulk of the advertisements from your digital visual landscape, it is time for us to turn our attention to social media. Reclaim control over your feed and use your social media power to curate your own personalised visual consumption – a consumption that supports you mentally, ideologically and emotionally. If you don't take control of your social media feed then your social media feed will control you. If you don't consciously decide what you want to see, the social media algorithms will decide what you see and your visual consumption will become even more narrow and even more impoverished. It's time to curate our social media feed so that it serves us and is a positive force in our life rather than a source of anxiety or just a colossal waste of time. Personalise your feed purposefully and consume the visuals that motivate you, inspire you and energise you rather than the ones that make you want to buy stuff you don't need or compare yourself to some impossible standard of beauty or perfection. When we slow our scroll and take more time over an image or when we actively like or share a post, we are effectively giving each visual a vote in some way, prompting the algorithm to show us more similar content and pushing the visual message to other people. Why would we vote for something that makes us feel bad?

Inspirational Feeds

While writing this book I was curious to know what others had on their social media feeds, so I asked several remarkable people about how they approach their visual consumption. I asked them about the accounts that were most inspiring to them and which artists and creators stood out to them. Mark Maclaine, a British educator, music producer and founder of Tutorfair told me that he follows the accounts of artists like Bansky and Invader. He also loves to follow a number of movie channels that share behind-the-scenes footage and clips of amazing movies.

I also wanted to find out if they had deliberately curated their visual consumption on social media. Very few had done so. The only person I interviewed who was deeply conscious of it was, ironically, an artist.

Jennifer Abessira, a renowned Israeli photographer, says that she is surrounding herself all the time with visual content that inspires her – even she is finding it hard to balance her visual content between the one that she knows she loves and the one she wishes to discover on social media. The artist is trying to stay open as much as she can by typing key words that are contrary to what she would want to see or wandering off to visual worlds that she may not be attracted to.

What I have found really interesting when having these conversations is that despite their awareness of this problem, the execution has been really challenging for many, even artists who are heavily immersed in the world of visuals. It goes back to the addiction we have towards a type of visual content that can be harmful to us. A movie producer I know is known for his left-wing political views and yet secretly mentioned to me that he loves to watch the TV show *Made in Chelsea*, which highlights the life of the young and wealthy in London, far away from his own political agenda. Visuals pose paradoxes and it's important to understand why we are attracted to these paradoxes. It tells us a lot about ourselves and that's why sometimes we are embarrassed to declare part of ourselves rather than embracing our own visual contradictions. Every individual shared with me the struggles they had in curating visuals that were kinder to their mental health. All of their struggles had one thing in common – they were all becoming emotionally hooked on visuals that made them feel worse. I call this the Ex Effect; named after that feeling of being drawn to checking on your ex's profile on social media when you are at your lowest (or at least pre my long-term relationship, that's what I used to do) to try and cheer yourself up, but inadvertently this is almost guaranteed to really make you feel crap. We often feel drawn to listening to sad songs when we're down and we engage with visuals in the same way – we are drawn to the ones that make us feel down when we are down. While we are conscious of doing so with music, we are often unconscious of doing the same with visuals.

Learning to change our habits is not about attaining perfection. I still am not perfect. I like to think about social media consumption as being comparable to junk food.

There's a brilliant book called *Ultra-Processed People* by Chris van Tulleken that exposes how the $700 billion snack industry capitalises on food-numbing, or numbing our emotions by consuming processed foods.[5] I'm someone that food-numbs; sometimes if I'm a bit stressed about something or emotionally triggered, I'll just go through maybe a packet of crisps or biscuits in two seconds in an attempt (usually failed) to numb my feelings. I quite like the analogy of numbing your feelings because I definitely do that.

I think you can numb your feelings with visuals in the same way. If you're having a really bad day, you're going to go for a visual programme that's easy to consume. It could be bingeing on show after show on YouTube or Netflix, or it could be spending hours on social media. I'm not judging you for your visual junk food habit, but I think it's an act of emotional numbing. So I think my advice for unlearning emotional numbing is being aware that we are doing it. It doesn't mean you can't go through a packet of biscuits or spend a rainy day watching comfort films; it just means being aware that if you're doing it regularly, it might mean you're trying to numb emotionally something that you should be processing. And visuals will be acting in the same way for you.

If you find yourself in an emotional numbing cycle, a visual detox can help you break the habit and regain control over how you choose to engage with visuals. It is also worth saying that in the very dense visual landscape and the visual world we live in, we have very little time to process emotions now because we are so continuously visually stimulated – your visual detox can help you reduce the visual noise in your life, both online and offline, to help give you more space to think and reflect.

Consider What Bothers You

Let's go back to your visual audit and take stock of what has a negative impact on your life online. For example, maybe it bothers you that men and women are so highly sexualised online? If it does, then actively stop following, liking and sharing sexualised content. If it appears on your feed, scroll past it quickly. Don't vote for them. This is easier said than done, but the more you make this conscious effort to skip past attention-grabbing content that you'd rather avoid, the more you will train the algorithm to show you different kinds of content.

Once you understand what kind of content bothers you, you start to notice it everywhere. I remember being in the Westfield shopping mall in London a couple of weeks ago and most adverts on display were sexualised: from the sexy woman with bright red lips promoting ice cream to the guy leaning erotically on the car he was selling. Did you ever pause and think: that's totally absurd?! It feels particularly tone deaf when the *Wall Street Journal* reported that 7% of American adults had sex only once or twice in 2022, 10% didn't have any sex in 2022, 36% once a month and the remaining percentage (53%) weekly – sex is not as big of a part of our daily lives as these ads would make us believe.[6]

It used to be that it was only women who were judged for their looks and hyper-sexualised in the images most of us consume, but that has changed. Men are now inundated with the same messages of inadequacy all the time, usually linked to a product that will allegedly solve the problem. The reason these adverts and this content are still being used to sell products is because it still works. The only reason it works is because millions of us are still following, sharing, liking or stopping the scroll to engage with posts that deliver that type of content. It follows, therefore, that if we don't like it or don't endorse it then we should stop following, sharing, liking or engaging with it. If we all did that then brands would seek to sell products in a different way and society might begin to follow suit. If the content stops capturing our attention then advertisers will alter the commercial visuals they're showing us.

Whether it is sexualised content, or whatever other category of content – do this process for every topic or type of post that 'bothers' you. For example, if you feel like you are seeing a lot of content pushing a materialistic definition of success, seek out media that shows you counternarratives about success. Success, wealth, happiness all come in endless shapes and sizes. Create your own definition and use it to guide what kind of content you allow into your feed.

Do You Want More Diversity?

Social media has taught corporations that they can no longer reach audiences by only showing us the same faces over and over again. The models and role models that brands are using have had to change to reflect the diversity of their audience. Those that don't make those efforts are being, rightly so, left behind. We don't live in a mono-culture anymore. People from all different backgrounds, ethnicities and sexual orientations want to be seen and represented in the visual landscape. And brands know it. Our endorsement of products and content effectively elects people to represent us. When we say, 'This person is really inspiring,' through our online engagement and that sentiment is shared by thousands more, then suddenly a different visual is on the billboard as we see our desires and votes expressed in society. The latest statistic about diversity in advertising, published by *Campaign* magazine in June 2023, stated that 27% of adverts that were integrating more diverse types of people had been outperforming the rest.[7]

The way you engage with visuals online can significantly increase the visibility of those messages. I had led my own company for eight years by the time the official portrait of ex-United Kingdom Prime Minister Theresa May by Saied Dai was revealed in September 2023, depicting her as a new kind of female leader. This visual felt so foreign and uncomfortable to me because it is so unsual to see women depicted outside of society's expectations. She isn't wearing a dress or your eighties-boss high heels, and her expression isn't clear either. It's cold and androgynous. It made me think about how, in the 16th century, Queen Elizabeth I was painted with a large dress and sumptuous jewellery, the subtext being that women couldn't be powerful through their bodies alone. Now Theresa appears militant with her army coat, garbed in a symbol of the patriarchy and yet reserved, as if still hiding from herself. It's also a portrait of a woman who is sixty-six years old in an era when society rarely creates images representing middle-aged and older women (most of our advertising shows young girls in their late teens or early twenties). The fact that I don't know what to expect of this visual shows how much room there is for so many people to be represented visually. I can't even begin to imagine the difficulty of minority groups in this context of visual representation.

Exercise
Focus on Your Feelings

While doing a visual audit on your social media feed, play close attention to how different visuals make you feel and when you are likely to encounter them. If you often scroll during your leisure time looking for a mood boost, pay attention to what kind of content brings you down and actively block, mute and unfollow as much of it as possible. If you want more feel-good visuals on your feed, start following visually expansive institutions, photographers, sculptors and artists that consistently share inspiring content.

If this sounds too simplistic, remember that this doesn't mean burying your head in the sand and avoiding all bad news or difficult subjects, it's about being intentional when you consume visuals that will impact your mood and curating your passive consumption to be more uplifting.

Photograph of brown wooden stairs with white concrete statues in
Vienna, Austria, by Reisetopia, © Reisetopia/Unsplash (7 May 2020)

Photograph of statue of Robert E. Lee on horseback on Monument Avenue in Richmond, VA, USA, by Amy Sparwasser. This sculpture was the focal point of Black Lives Matter demonstrations in 2020. © Amy Sparwasser/iStock (6 July 2021)

'Person kneeling before police officers' photograph taken in
Washington DC, USA, by Spenser H, © spenserh/Unsplash
(22 January 2017)

'Cheerful' photograph by Derrick O. Boateng,
© Derrick O. Boateng (2022)

'Queer Voices at Kew Gardens' immersive space designed by
Adam Nathaniel Furman for a film-based installation featuring
interviews with horticulturalists, scientists, authors, drag artists
and activists as part of the *Queer Nature* exhibit for Kew Gardens,
© Adam Nathaniel Furman (September 2023)

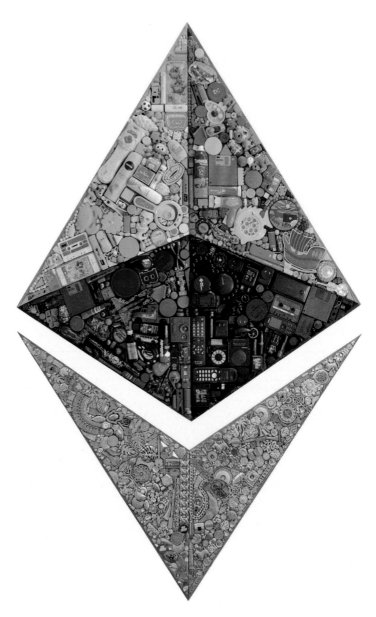

'Abstraction Materialized II' mixed-media artwork
by Elisa Insua, © Elisa Insua (2023)

'Black Madonna' AI artwork created using MidJourney
by Delphine Diallo, © Delphine Diallo (2023)

'Hope' photograph by Claire Luxton, © Claire Luxton (2019)

Use your social media power by consciously deciding who and what you follow, as well as what you like and share, to vote for the people, ideas, and the type of society you want to live in. Remember that the algorithms are smart so they are also paying attention to when you stop scrolling and how long you look at something. So, it's not just about what you do actively, it's also about measuring what you do passively to determine what is served up to you in your feed.

If this irritates you – it should! Consumers have more power today in vocalising or calling out brands. You can't just hide behind a beautiful advert and not be up to standard. I believe that the relationship and the balance of power between us and image creators has changed. Social media is part of it, 'cancel culture' is part of it, but also the fact that now even the media reports differently about brands. We are all much more engaged in questioning the messages we're receiving and we can all learn to engage critically with what we see and think about the visual messages we're consuming and consciously bring in more diverse perspectives by actively seeking them out and supporting them.

It's hard to learn to control which visuals to interact with; control calls for intentionality. Setting timers and screen time limits on different apps can help nudge you out of passive consumption and help keep you from being sucked in. Being aware that the first image you look at when you wake up will likely affect how you experience your day can help you to be more intentional about not looking at a screen right away. You could introduce a new habit such as starting your day with a nice walk in the two streets around your home to ensure that you aren't bombarded with visual messages right away. I think being more intentional about what you want to see first thing in the morning and resisting the urge to switch on your phone and look at whatever comes up sets the intention that ends up guiding your day.

Even when we do see ourselves represented in a lot of the visual messages we receive, diversity in our visual consumption is still beneficial to our wellbeing. I believe that we are not hearing a diversity of visual narratives, which is why we don't have a diversity of visuals. I also think this is why we don't feel great – we are being constantly pushed at with a few oppressive narratives, one of which is the very commercial one mentioned in earlier pages. The solution to this is to bring more diverse visual narratives into our lives from a wide range of creators and types of people. This serves to expand our understanding of the world, discover new ideas and expand the possibilities for how we can express ourselves.

Online Friendships

One thing that makes our social media feeds difficult to curate is our friendships. These are social channels and perhaps some of the friends you love the most are the ones posting the content you get triggered by all the time. Thankfully, there are some strategies to help you get some digital distance. Every friendship is different, but think about trying these techniques:

- Mute their content so that you don't see it on your feed or get notifications. For some friendships this might be enough! In most cases, they will be none the wiser and you can maintain your online connection with that person.

- In addition to muting your friend, set aside intentional moments to check on them and manually check their profiles for updates.
- In some friendships, especially with close friends that might expect you to see their updates and get upset if you don't engage with them online, talk to them about how you are limiting your social media use for your own wellbeing or are avoiding certain kinds of content (for example, if you are trying to conceive a child, seeing unexpected images of your friend's baby might be difficult for you emotionally). I recommend that you lead with how you're taking steps to manage your own feelings and wellbeing rather than criticising the visuals themselves, to help them understand your point of view and know that you still care about them even if you are not watching their daily updates.

Be Your Own Moderator

Of course, social media curation is about more than just your feed. All social platforms have to do a certain amount of moderation to prevent unwanted content from taking over, and many of them have tools that invite users to be a part of the process through blocking and reporting content. Understanding how content moderation works can help you better recognise how it affects your digital visual landscape.

I am very anxious by nature, and so I have an array of toolkits to deal with my anxiety. I am very vigilant and very sensitive to triggers around loads of different things (not just my visual environment). While I function highly as a person through that anxiety, I acknowledge my anxiety and have to be extra conscious of the content that would make me extra anxious.

Empowerment for me, therefore, is where understanding what triggers me – what makes me feel insecure or anxious or uneasy – is key to determining and comprehending what harmful content is like for me. It can be the same process of content moderation for you.

Because I know what content is harmful to me, I monitor my content and eliminate it. And I think that act of monitoring is really the one action that we all need to be doing for ourselves. Be proactive about blocking or muting things that show up on your feed and disrupt your day. While it is important to interact with content that can be upsetting, in order to be a well-informed person, you can choose when and how to interact with that content instead of getting blindsided by it while passively browsing or searching for other kinds of content on your feed. This individual moderation is a microcosm of what websites and media platforms attempt to do on a bigger scale to ensure that the algorithms are showing people appropriate content.

Let's take the social media platform TikTok and look at it from a 'content moderation' perspective as an example. Overall, TikTok's unique format and user-driven content are creating a new and innovative space with the capacity to be more accessible, diverse, educational and engaging than traditional educational platforms. There are several things it does well:

- Bite-sized learning: TikTok's short-form video format allows for easy consumption of content in small, manageable chunks. This format can be particularly effective in an educational context for younger learners who have shorter attention spans and prefer visually engaging content.
- Accessibility: TikTok's vast user base and algorithm-driven content distribution mean that content can reach a wider audience than traditional educational platforms. Users can easily discover and access educational content on TikTok without needing to actively seek it out.
- Diversity of content: TikTok's user-generated content model means that the content available on the platform is diverse and varied. Users can find content on a wide range of subjects, from science and maths to art and culture.
- Creative expression: TikTok's format also encourages users to express their creativity and share their own knowledge and skills with others. This creates a collaborative and dynamic learning environment where users can learn from each other and build on each other's ideas.
- Informal learning: TikTok's emphasis on user-generated content and informal communication

means that users can engage with information in a more relaxed and fun environment. This can help to remove the barriers to learning and make educational content on the platform more engaging and enjoyable.

However, not all of TikTok's content is educational and the algorithm is not necessarily pushing educational content. TikTok, like other social media platforms, allows users to upload whatever they want and then initially relies on algorithms to identify content that possibly violates their guidelines, using human moderators to pass judgement on content that the algorithm flags as a potential violation or appeals from users who have had their content taken down by the algorithm. Content flagged by users also gets escalated to these human moderators.

At the end of the day, content moderation boils down to individual judgement calls about whether a visual has broken one of the platform's rules. It is informed by a company's policies, legal requirements and prevailing cultural norms. It is not a perfect or foolproof system – sometimes harmful material slips through the precautions that are put in place and sometimes content is blocked or removed unfairly.

What Does 'Harmful Content' Look Like to You?

The major issue with content moderation is the determination of what harmful content looks like – and how to moderate an algorithm without discriminating against certain communities. It is a difficult and nuanced task that is time-consuming and expensive to do well, which is one of the reasons that even a well-intentioned and well-designed moderation system will miss some content or block content that might not actually be harmful.

Enter *participatory content moderation*, a new model which provides the user with the opportunity to moderate and govern the content shown on an app or website, through a layered system of tagging. This model has most recently been picked up by Tumblr, who have reintroduced nudity to their website this year, on the condition participants use the 'community labels' so that people can choose which content they want to see.[8] This has also been tested by the company Reliabl, who are using tags to filter content for people which is specific for them whilst still excluding exploitative content. They claim the current content moderation system 'fail[s] to incorporate accurate data on marginalised communities, and prevent[s] user participation in the data classification process – resulting in unreliable unsafe algorithms.'[9] Introducing data models in collaboration with users can help construct a visual consumption for participants which is both more inclusive and makes them feel seen, without the repercussions of being restricted or made invisible.

While at face value participatory content moderation may seem a daunting task, what it requires of the participant is uploading a photograph and using tags, such as #art #nft #womenempowerment #queer, to match the image with other works in that context. By doing so, it helps viewers find what they want to see more easily. The difference in this hashtagging on Tumblr compared to hashtagging on a platform like Instagram, in this respect, is Tumblr's content moderation promotes participation, allowing viewers to add or modify tags, unlike on Instagram, where the tags are created by the initial poster. This gives the algorithm a broader view of what the content actually includes.

Most social media platforms have tools that allow you to mark content as not appropriate for your feed, block or mute certain subjects and users or install 'parental controls' to limit more sexual or violent content. Intentionally using these tools regularly will improve your own feeds and help the algorithm better understand what content other people might also want to see less of.

Shadow Banning

While moderation is essential to ensure that we aren't bombarded with harmful content, sometimes content moderation can reproduce discrimination against marginalised groups. This can be due to grey areas within their content policies or the biases of the moderators. One of the strategies companies use to subtly moderate their content is 'shadow banning' – hiding a user's content from non-followers without making them aware that this has happened. Shadow banning is not an official policy for most companies, making it notoriously hard to prove, but there have been several high-profile cases that have demonstrated that it is disproportionately being used to hide content from minority groups.

In 2019, a number of women took to their Instagram profile settings and swapped their gender to male, due to the media conglomerate 'shadow banning' women whose skin is exposed, while for men the same rules did not apply. The most targeted of these groups were women pole dancers, whose popular hashtags, such as #polefitness and #femalefitness had been completely censored by the app in 2019, while poignantly #malefitness remained active. As a result of changing their gender on the app, these same pictures were no longer hidden, proving that the algorithm was hiding their posts because of their gender.[10]

This feels reminiscent of 19th-century female writers like Mary Anne Evans, better known for her male pen name George Eliot, or the Brontë sisters, aka Currer, Ellis and Acton Bell, who needed to pretend to be male in order to be successful or to be published at all. Similar instances of shadow banning have also applied to the LGBTQ+ community on the app, who use hashtags such as #bi and #iamgay, and more broadly it has been documented that all the major media conglomerates' content moderation, including Twitter (X), Facebook and TikTok, has led to discrimination of minorities.[11] In December 2022 Instagram added a feature for professional users, which would let them know when their content had been deemed 'not eligible' to be recommended to other users. However, there is still a large number of people, particularly Black creators, sex educators, fat activists and drag performers who are regularly confronted with the problem of dwindling engagement numbers and who are left wondering if it is their content or in fact shadow banning that is affecting their reach.

This phenomenon is not limited to Instagram. It happens across most social media platforms. For example, there have also been repeated instances when Black TikTok users have documented that after posting content related to the Black Lives Matter movement there were noticeable declines in their viewership and engagement on their videos, or that TikTok's community guidelines were not being fairly applied to Black creators.

In short, shadow banning is not just about gender and race – it can be a form of silencing groups of people in a type of visual. This is because visibility is power. Being visible and represented is powerful and being shadow banned removes that power. It's like a bloodless coup d'etat. If there's only a certain type of people being represented, the others feel they don't exist visually. And that echoes my earlier point about the average-height man having a city visually designed around his dimensions, which is ultimately an act of constant silencing. So this is why shadow banning matters as a discussion: it's the act of silencing a type of people. As an individual user, being aware of whose content you're seeing and making sure to engage with a wide range of content can help show the algorithm and the platform that you want to hear their voices. If you're repeatedly not seeing certain types of people or subjects that you're interested in, it may be a cue to seek them out in other forms of media where their visual narratives can reach you more freely.

A Brave New World

There is a shift taking place in which the digital spaces we inhabit are playing an increasingly important role in our lives. We can see this manifested in a shift from physical art to digital art. In just the past few years, we've seen the introduction of NFTs, which allow digital art to be bought and sold in a whole new way, and the birth of generative AI, which is when AI programs create entirely new visuals based on what they've learned from existing images and video clips. New developments in virtual reality promise that new ways of interacting with digital visuals are right around the corner. The digital spaces that we inhabit are changing rapidly with the advent of new technologies, which makes it more important than ever that we invest in our digital literacy and remain aware of new kinds of visuals entering our landscape.

As digital visual creation becomes more sophisticated, our ability to create artificial images that look real increases. There are certain things that instinctively tip us off when we're looking at an image. The perceived 'realness' of a visual is often determined by colour saturation and contextualisation. Full colour is viewed as more real than black and white. Deep perspective and full illumination of light and shade are also considered more real than their opposite. Fully conceived backgrounds with natural shadows are more real – this is why fake Zoom meeting backdrops feel so 'off'. We can tell when the light in an image is not casting realistic shadows. When we know an image is fictional, such as when we're seeing special effects in a movie, the surreality doesn't usually bother us, but when something looks a bit off and we are expecting an image of reality, we can feel uneasy without knowing why – this is your brain picking up cues that the image doesn't match reality.

Sometimes there are visuals that blur the line and make it hard for us to tell the difference. For example, the background of the Microsoft Windows screensaver is a landscape – green, blue, clouds. Just through the proliferation of Windows on computer screens, this is probably one of the most viewed images in the world. It was captured by a Californian photographer who was randomly passing the scene in his car, and shot it. This landscape is called Bliss. And what's even more interesting is if you visualise it right now, it is actually the least representative of nature, of a natural landscape, you could get. There's almost no life in it. Nature doesn't really look still like this with blocks of colours. It's usually much more complex. So this, the most perfect idea of a landscape, is actually the least likely to exist in nature. When we view the image, it feels very saturated and oversimplified so we don't read it as 'real', however, in this case it actually is a real photograph.

As generative AIs become more sophisticated and we are debating the copyright issues around how the people who created the visuals they are trained on get compensated, we need to keep in mind that at the end of the day AI represents a set of new technological tools that are here to stay. Like all tools, we pick and choose the most appropriate among them for our purposes and needs and there will be ways to use these tools ethically.

Case Study:
Ellie Pritts

AI has the potential to help more people create and share their visual narratives. I have an artist, for instance, Ellie Pritts, who has a hereditary condition called CMT disease which causes nerve damage and impacts the muscles – she used to be a painter but can't paint anymore. AI has been enabling her to continue to be an artist and further her artistic vision. She says that 'collaborating' with AI has opened up a world of endless possibilities for her and helped her to love herself again and continue creating in ways she thought she couldn't – opening up opportunities to work on physical sculptures and fashion. Her AI has fostered new creative connections with people all over the world. She trains AI on the catalogue of visuals and written words she has already created in order to collaborate with her past work via AI to generate new visual ideas that build on her previous ones. She says, 'It's such a joy to breathe new life into old dreams and watch many of them become more real than they ever have been. With AI I am the only thing standing in the way of my dreams, not my circumstances.' Ellie Pritts demonstrates that AI can be a tool that extends human creativity.

Another new development in digital art is that we now have new tools for visual ownership via NFTs. NFTs, or Non-Fungible Tokens, are digital assets that can come in the form of art, music, in-game items, videos, and more. They are bought and sold online, frequently with cryptocurrency. What sets them apart from other digital assets is that they are encrypted with blockchains that certify the ownership of the visual.[12] Some of the NFT projects that you might have heard of are CryptoPunks, and later, Bored Ape Yacht Club, a series of 10,000 images of apes that are all unique. The first recorded NFT was 'Quantum', created by Kevin McCoy and Anil Dash in May 2014. It consists of a short video clip made by McCoy's wife. 'Quantum' was the first demonstration of 'monetized graphics' and was registered on the Namecoin blockchain. McCoy sold it to Dash for $4.[13] A bargain considering a version of that original NFT was sold by Sotheby's in 2021 for $1.4 million.[14] This technology was improved with something called ERC-721 standard which provides basic functionality to track and transfer NFTs.[15] ERC-721 allows someone to create a unique piece of digital art and cryptographically sign that image with a unique key that verifies its authencity. This means that, for the first time, the image only exists in one form and it is easy to identify copies of that image and distinguish them from the 'real' virtual object. The ability to create digital pieces of art that cannot be copied has forever changed how we think about and value digital art. In 2021 Beeple caused a sensation when an NFT of his artwork sold at Christies for $69 million.[16] A year later the digital art NFT market ballooned to $41 billion, out of thin air.[17] The conventional art market was worth about \$50 billion in 2022.[18] Just think about that for a moment. A market that didn't even exist a decade ago is almost worth as much as a market that has been around for centuries. As I am writing this in 2023, the NFT market has seen a collapse in its overall value, but the rise of digital art more generally feels inescapable and this market will continue to grow over time.

Another new development in digital visuals is the push to create an entirely virtual world, referred to as 'the metaverse'. Let me demonstrate for you what that might be through a couple of examples you may know of. The term 'metaverse' came from a novel by Neal Stephenson called *Snow Crash*.[19] In the novel the metaverse was originally meant to be a space free of commercial interests but in the end became just as commercial as the 'real' world, recreating a lot of the same inequalities that exist in the 'real' world of the novel. The irony is that attempts to create a real-life metaverse in our world are very much driven by commercial interests and large corporations. The 2018 film *Ready Player One* by Steven Spielberg, based on the book of the same name by Ernest Cline, shows us another vision of how the metaverse could become harmful: in the movie, people just hate their current world, so they constantly escape into the metaverse. It suggests that there's a danger of making something that is so perfectionist that it will make us hate human life and hate being in the real world. While these science-fiction-based worlds do not exist entirely today, there are hundreds of millions of people every week spending time in video game worlds, such as

Fortnite's world of Reality Zero which has over 500 million users who spend on average six to ten hours in it per week.

What do we need to consider about engaging with or even building such virtual spaces? Speilberg's film warns us that the way we build the metaverse is going to affect how we look at our physical world. So, for me, that should take us back to nuance and moderation. If you become completely absorbed by a new visual space, the problem is that you will struggle to reset your reality, your visual reality. Don't misunderstand me: the visual digital world is fine as long as you can reengage in your current visual reality, and that is the part that is becoming harder and harder as we are already spending more and more time in digital visual spaces.

Between NFTs and the metaverse, we can see that we are on the verge of entering a new visual landscape: an immersive online space that people can inhabit, one that will offer new opportunities for artists and for people to explore new realities. They can create new communities and bring people together around various interests and causes. There's a lot of positive potential in these developments, but these spaces are also creating new places for us to encounter visuals that can impact our wellbeing, so it is important to reflect on how consuming them impacts us as we bring them into our lives.

I believe that our digital visual environment is having a substantial impact on what we believe we need, how we see ourselves (in a positive or negative light) and how, in the end, we interact with others and the world. By doing visual audits and detoxing, we can control more aspects of our lives and happiness than we may have realised possible, giving ourselves the space to step back and think critically about the visual messages we are absorbing.

I believe the explosion of social media has warped people into believing they need more than they actually do to be happy. We all need money to survive but the visuals we consume tell us we must become super wealthy to be considered successful and happy, and absorbing these visual messages makes it harder to be content with what we have. In 2011 in Daniel Kahneman's landmark book, *Thinking Fast and Slow* he included his 2010 research that showed that there are no money-related happiness gains in the US beyond the $60-$90,000 a year income band.[20] His argument was that there is a point at which more money doesn't improve your happiness. In 2023 a researcher named Matthew Killingsworth worked with Kahneman to replicate the study and they found that that point, where money doesn't increase happiness (which when adjusted for inflation in 2023 was $100,000) is where people are experiencing miseries that are not improved by higher incomes, such as bereavement or heartbreak.[21] These studies show that while an increase in income in the short term could temporarily increase your happiness, in the long term you stop seeing a boost in happiness due solely to a higher income. Beyond a certain threshold, money stops significantly alleviating the things that can make you chronically unhappy and your stress levels, health and wellbeing and relationships have a greater impact on your happiness than income alone.

According to Martin Seligman, the father of positive psychology, what really matters for happiness is satisfying work, avoiding negative events as much as possible, being in a healthy intimate relationship and having strong social networks. Yet our visual landscapes, mostly through digital visuals such as advertising and social media feeds and posts, are artificially making people believe they need more, and in doing so making it harder for people to be satisfied or content with life. Use your detox to push back against these visual messages and take back your life and happiness.

Your Visual Detox Plan for Your Digital Presence

For many of us, a significant part of our visual consumption takes place online, where we don't just consume images, we also project a narrative about ourselves through what we post, share and publish. For some people, their only real presence online is social media but if you have a business or an online presence linked to your career, you also need to consider how you are represented through your website and any other mediums that present your professional self.

To audit your digital presence, turn back to chapter 1 to evaluate the visual narrative that you want to build in your life and consider what you want to say about yourself online. Evaluate all of your platforms: what could someone see about you online right now? What judgements are they likely to make based on those visuals? Make sure you are happy and comfortable with those likely judgements, or make some online changes to alter the visual narrative. Consider the messages you are sending through the visuals you are presenting. This is also true in our professional social media sites such as LinkedIn, websites and social media feeds.

Think about the type and quality of the images you are using to represent yourself and who will likely be seeing them. Are they showing you the way you want to be seen? I am always surprised at how many people include photographs of themselves on their professional profiles and websites that they clearly took themselves. If that's you – get a professional shot taken and change that photograph today. If this option is too expensive, YouTube has a myriad of videos you can watch to train yourself to take better portraits.

When it comes to professional websites, pay attention to the words you are using too. If you are not a writer, hire one. Get the content saying what you wish, how you wish, then get some outside feedback about 'how it reads'. Does it still reveal more of your personal life than you'd like? If so, have that content changed.

Chapter 8.

Re-imaging Your Visual Landscape

The easiest place that you can control which visuals you see is within your home. As a child, I wasn't allowed to change much of my bedroom and I didn't like its aesthetics, but the second I got my first flat, a small space of about 10 squares metres (it was a bedroom more than a flat as we were sharing a bathroom and I had a little stove next to my bed) at my French university in Poitiers, I discovered that making my 'home' visually appealing was more important than anything. I became the queen of making any space look amazing with very little and the happiness it gave has been tangible since I was seventeen years old. Even if you are renting a space, there are always small things you can do to improve the visuals you see every day. It could be a change of curtains, a few drawings on the walls or a vase with a couple of fresh flowers. Once you start, you will see a difference.

Theories about the psychology of interior design have been around for thousands of years, so there is a lot of information out there to help you learn how to make your space work better for you. Indian Vastu Shastra and Chinese Feng Shui are perhaps the best-known examples of ancient approaches to architecture and design, meant to create a sense of harmony between nature, architecture and the inhabitants of a space. With modern research tools neuroscientists are now able to demonstrate that the impacts of interior design elements are more than theoretical – they evoke tangible positive or negative emotional responses in people. Research suggests that creating a visually pleasing home environment can lead to improved mood, reduced stress levels and increased happiness.

For example, a study by Graham, Gosling, & Travis (2015) identified that the organisation of a room is a key component to consider when designing spaces that support peace and mindfulness. The study revealed that interior designs that maximise the positive effect on mood were all arranged in a way that fostered social interaction and were open and easy to manoeuvre.[1] We also know that colour can impact how we feel too, and last but by no means least, the positive impact of art on our wellbeing does not just apply in the context of museums or other art exhibitions. Just as patients in hospitals benefit from being exposed to art, having art in your home will also have a positive impact. According to Professor Semir Zeki's research at UCL, viewing art gives the same pleasure as falling in love – he found that looking at art immediately stimulates the same parts of our brains that trigger the dopamine pleasure response.[2]

That said, even with all the research out there, taking control of your visual narrative in your home does not require an interior design education or deep knowledge of the science. You can start small by reflecting on what feels good and learning from your visual audit. When I asked my sister about whether she designs her home she told me, 'I'd love to but I live in a terrible place and want to move as fast as possible. I don't try to decorate it at all because it would mean I might stay longer. I am in denial! But I do still love to create nice visuals of memories and pictures on my walls so I feel more at home.' Regardless of our situation we can all do something to improve our visual consumption at home, and it is worth prioritising. Everyone needs a place in their home that makes them feel safe and comfortable. You deserve to create a space for yourself that meets your psychological needs. Psychological comfort isn't a luxury, it's a necessity. Use the suggestions below or come up with new ideas to make your home your comfort zone – a place of peace and renewal.

Having come from a modest background myself, I strongly believe that all of us can create a powerful visual environment that empowers us, and art can do a lot to help us achieve this. The way I look at the art in my home feels like a visual conversation I have created; it's like when you invite your favourite people to dinner and the conversation flows easily. Each piece engages me at different times of the day. I have one called *The Planets* by artist David Aiu Servan Schreiber which consists of two large paintings with so many different shades of blue and a depth of texture that result from his way of literally painting with fire (he burns the top coat of the painting creating layers of painted cracks on the wooden panel, it's incredible) that brings me peace whenever I need it. I just have to sit next to it to feel calmer, my eyes can literally get lost in the texture and the blues absorb me. In another part of my home the fierce photographs of artist Delphine Diallo speak directly at me, as I lift my eyes up to them, to shout, 'Be brave! Keep going!' The visuals in your home don't have to be expensive, they just have to speak to you.

Your Visual Detox Plan for Your Home

Transforming the visual landscape in your home starts with a plan. Consider the following suggestions and action what feels right for you and your budget. Not all of the suggestions will work for every space, but together they support a sense of wellbeing and calmness in the home.

Even if you have decorated your space in the past, it is worth taking a moment every once in a while to update your visual landscape at home to suit you better. German American psychologist and psychoanalyst Erik Erikson developed a theory of the Stages of Psychosocial Development in the 1950s that describes the different stages of development that we all go through.[3] Each stage from child to adult into old age reflects who we are 'at that time'. We often see this through how we dress but it is also reflected in our surroundings at home. Looking at your home now – is it stuck in a stage of your life that is more reflective of who you used to be rather than who you are now or who you aspire to become?

Consider each room as an individual canvas. What is the purpose of the room? What do you do in that room? How often do you use that room? How long do you spend in the room? Start curating your visual consumption at home by focusing on the rooms you spend the most time in or enjoy the most. Things to consider include:

- Colour
- Light

- Clutter
- Nature
- Textiles
- Walls

Colour

As we saw in the colour section in the first half of the book (page 91), the colours we choose trigger different emotional responses and associations. The research in this area is still relatively new and more studies are needed to fully understand the effects of colours in our homes on our mood and mental health, but it does indicate that we should take choosing colours seriously.

Most of us have a favourite colour. Mine is blue – it reminds me of all the hours I spent staring out to the ocean as a child. I certainly find it soothing in the same way I found the ocean soothing. As an adult, blue is a deliberate addition to almost all my visual spaces at work and at home.

Colour is incredibly meaningful and yet often, especially at home, we are encouraged to choose neutral tones. Why? If you have a favourite colour that makes you happy – use it. If you love bright, bold colours then consider creating a feature wall instead of painting the entire room, or painting furniture in that colour to bring it into the space. This will give you the visual benefit of the colour without the intensity of painting a whole room. Temporary wallpaper can also be much easier if you cannot paint on your walls or you can also incorporate your favourite colour through curtains, art, ornaments or bedding. You can start with little touches and build it up over the years as you discover which colours have a positive impact on your life. If you're not sure which colours to use, refer back to the colour section and your visual audit to find colours that inspire the energy you want your space to help you access.

Light

Creating a home that supports your visual experience is profoundly unique. What we like and don't like is individual and based on a myriad of contributing factors. One aspect of home that is largely universal is light. Most of us feel better and more engaged when we can enjoy natural light in our home. Colour too will play a role here. If our home does not receive much natural light then using dark colours can make that worse – consider white or light colours for these spaces to reflect the light you do have. Mirrors don't just amplify and reflect the light but they will make the room look bigger too and can be a useful tool in maximising the light in your space.

In 2016, researchers Alan Cheung and Richard Slavin published a particularly interesting study on light and blue-enriching light exposure.[4] It shows that both cognitive function and sleep quality benefit from good exposure to daylight. As for the effects of blue-light exposure on mental health, they suggest that prolonged exposure to blue light from electronic devices such as smartphones and tablets may be linked to depression and other negative mental health outcomes. Researchers from Harvard Medical School and the University of Toronto have shown that blue light is a potent suppressor of melatonin, which your body uses to set your circadian rhythm and help you get good sleep. Lack of sleep can contribute to many negative health outcomes, including cancer, diabetes, heart disease and obesity.[5]

Take this into account when you decide where you want to keep your electronics and how you arrange your furniture. If you struggle to sleep or you don't get the sleep quality you need, consider removing all forms of technology from your bedroom. That includes TVs and smartphones. Just try it for a week and see how you feel. Being aware of where you get the most light at different times of day can help you decide how you want to use your space. Good indoor lighting can also help to improve how you feel in your space. Use an array of lighting options to create a different ambience or mood in the room. Try bright overhead lighting in the morning as everyone is preparing for the day and softer, side lighting in the evening as everyone is winding down and preparing for bed to echo natural light.

Clutter

When Marie Kondo's book *The Life-Changing Magic of Tidying Up: the Japanese art of decluttering and organizing* was published in 2014, she very quickly became a household name.[6] Ever since Netflix produced the tie-in show *Tidying up with Marie Kondo* in 2019 there really has been no escaping her decluttering approach of only keeping things that spark joy and having a place for everything and everything in its place. Self-optimisers around the globe have since worked on reducing their unnecessary possessions. For those of us who find it harder to follow this regime, there was some consolation provided when in 2023 Marie Kondo declared in an interview with the *Washington Post* that since the birth of her third child she has 'kind of given up' on her method, focusing instead on spending her time in a way that was right for her at this stage of her life.[7] Still, there is no doubt that too much mess and clutter can have detrimental effects on your wellbeing.

Just as light is a universal benefit in any home, too much clutter is a sure-fire route to disharmony. When what you see the moment you wake up is mess and clutter, that visual is already negatively impacting your visual experience and altering your visual narrative. If you don't love, use or wear something, consider donating it to a charity shop. Either replace it with something you love and personalise your space deliberately to reflect who you are now, or don't replace it at all. For the things you love and use, make sure that they have their place and get everyone in the home to pull their weight to make sure that they stay in their place. We don't need to turn into show-home minimalists to have a comfortable visual landscape at home, we just need to pay attention to how our living choices impact our visual consumption and how we feel. Make adjustments if necessary.

Nature

Bringing nature into your home through plants and flowers has been shown to improve our mental health and brings a deeper connection with our home. Even a twenty-minute 'dose' of nature is enough to reduce our stress levels, and studies from around the world have shown that exposure to nature can help alleviate symptoms of depression and even reduce food cravings.[8] Flowers are especially lovely because they also look great. Thankfully, many supermarkets now sell beautiful flowers at a very reasonable price. As soon as Covid hit and we entered our first lockdown in the UK I organised for many of my team and artists to receive bouquets of flowers regularly. And I included myself in the order. It was a small delight in the midst of a challenge that improved our visual consumption and made us all feel better.

For a more permanent way to get these benefits, consider a few potted plants, easy-to-care-for green plants or even a window box of herbs. If you struggle to keep plants alive then there are some amazing fake plants that look so realistic it's almost impossible to tell and they will still lift your spirits and improve your visual consumption.

Textiles

While the research (and I know I say this often) is in its infancy about the psychological effects and benefits of textiles and textures in one's visual narrative, Albert Read makes an interesting point in his book, *The Imagination Muscle*. He writes that in a society where we are constantly stimulated, we are losing the time and space within our lives to let our imaginations run free. [9] His premise (and I agree!) is that imagination is key to psychological wellness. Our imagination – our way of seeing and existing visually and creatively – matters, and we should all contribute creatively to the world we live in. In other words, our creative imagination can be a problem-solving resource, and as it is, according to Read, a muscle, we strengthen our muscles by using them. I believe that seeing different textures in a space exercises that visual imagination muscle through the increase in complexity of what the eye sees, first, and then by creating an object of contemplation for a longer period of attention.

Are there different textures and textiles in your home? Are there spaces where you can add visual interest with texture? Perhaps textured wallpaper to add a bit of luxury or glamour, tapestries as wall coverings, or thick velvet curtains in the living room to block out the winter chill. Utilise different types of fabrics, opulent textiles and finishes throughout your home, especially your bedroom.

As it's the first thing you see when you wake up, make sure your bedroom is a visual feast. This is *your* room. Express yourself. Make it your own so that it makes you smile when you wake up every day. If you have any eclectic items that the family raises an eyebrow at – put them in your bedroom. Make it your sanctuary. Consider an ornate or creative headboard or lovely Egyptian cotton sheets, layered bedding, cosy throws, comfy pillows and luscious rugs. Luxurious bedding used to be the domain of the wealthy but it is now possible to create a really delicious sleeping experience without the eye-watering cost. Little things that can make a huge difference to our sleep quality and how we feel when we wake up.

Walls

Something that's recently been a subject of debate by a few architects is the way a blank wall is perceived. Is it so plain and devoid of detail for the eye that we just quickly bypass it? Or is it soothing so our eye stays on the blankness? Does it turn into something that's going to make you feel more anxious? Studies about this are really in their infancy but the core idea is that looking at a visual complexity – a painting in a museum with (typically) its overload of details means you spend longer in front of it to comprehend and appreciate it – versus the enigma of a blank wall that allows the mind to either freak out from inactivity and move on to something else or dive into the blankness in quiet contemplation is the issue they are looking into.

Any in-depth visual examination is better for the brain and better for you in the sense that you are engaged longer and in a more complex manner (such as reading a whole book) and you're able to absorb the content as the mind works through it. Consider which approach works to create the visual narrative you want in your personal spaces. Do you want spaces such as a blank wall that will give your mind breathing room or complex statements such as a painting that demands examination that will stimulate your mind? What could you add to your walls that would improve your visual enjoyment – either as part of the wall (paper, paint) or on the wall (art, tapestry hangings, etc.)?

Visually, we all have very different things that make us

feel happy. Some people will love maximalist spaces with flyers and books and loads of things hanging on their walls, and that makes them feel really happy. They might prefer to live in a tiny street or lane where the houses are on top of each other and there's loads of colours to them. Some people will want something that's much more decluttered and plain or even minimalist. And it's the same with artworks.

If you're not sure what to hang on your wall, try to remember something, an image that made you happy, and then try to comprehend why it is so. Use what we've just been through in all these pages. What is the composition of that image that made you happy? Maybe the focus of the composition was it for you, and this is a composition that you are really drawn to. Maybe it is quite a blurry image and you liked it because you could escape into the blur of it. Whatever it is that makes you happy, bring it into your space.

For most of us, art is something that belongs in country estates or galleries, an indulgence for the wealthy. And clearly that is true when we are talking about Picasso or Rembrandt, but there are thousands of local artists showing their work in coffee shops or visitor centres up and down the country. Visit charity shops, car boot sales or local markets and just look for art, in whatever form you appreciate. Your visual audit can help you identify the colours and subjects you're most drawn to to help you find the art that will help you shape your visual landscape at home. When I'm selecting art for myself or my clients, I always go back to the visual message first and find a piece that fits the visual narrative we want people to experience within the space.

Your Visual Detox Plan for Your Workspace

You may have less latitude for making modifications to your workspace, but depending on your job you may be able to make positive changes. In the workspace, it's the function first (e.g. a supermarket versus private offices with conference rooms), followed closely by who's in charge of that functionality – for any larger changes, you'll need to convince other stakeholders of your vision. Let's look at an example of this.

Start by expanding your awareness first. What kind of work environment are you in? Do you have a space that is 'yours', such as a desk or an office, or is the visual landscape dictated by someone else? I heard of a woman working for the first time in a huge city clothing thrift store; she saw that tops were just grouped by 'women', 'men' and 'kids' on long, long rows of clothing rails. She thought, 'There's no visual hook here. No visual leading a buyer to any of this!' She took it upon herself to reorganise each of those racks of clothes not just by size (as is usual, but not always visible from a distance) but by colour groupings. In her 'red' zone she placed reds, oranges and the vivid and pastel versions of those colours, and so on through the colour palette down the rail of tops. Her new co-workers loved it – it was a new visual – and they noticed they were selling more of them. Buyers gravitated first to 'their' colour. So, understanding what your work environment looks like and will allow, as she did, is a first step. What and who was it designed for (buyers looking for cheap tops)? What is the design supposed to achieve (make a quick sale)? This woman took the perspectives of clutter (and organised it), colour and texture and textiles (from the very products they were selling), and made a welcome improvement to the visual environment.

Examine your work area as you do any other visual from the approaches we've discussed. You could be thinking that the room feels very cold. The lines feel very straight, the space feels very functional, and it feels very stern or limited in terms of colours. Maybe everyone wears suits; maybe it's all-female, or contrarily, male-dominated (and what does that do visually for you)? If you can identify visual changes that will improve everyone's comfort and productivity, you can make a good argument for change to those in charge of decision-making.

You can also look for opportunities to change how you visualise information within your work. We now know that a lot of people are visual thinkers. You may have a majority of verbal thinkers in your company so, for instance, you might even rethink the visual environment based more on how you want to attract talent and a diversity of people. At L'Oreal the staff do a personality test that gives them each a colour. They use these colours internally, such as in email sign-offs that state, 'I'm blue and yellow,' or 'I'm red and green.' The idea is that everyone does this according to colours and what type of personalities they are to help email recipients better address or communicate with that personality. It helps everyone in a diverse environment to adapt to a wide range of personalities according to those visual cues.

A system like this allows management to conclude things like, 'We are missing loads of yellow and red in this company and yellow is quite creative. So maybe that's because actually, our work environment is not very creative looking.' I think this kind of visual analysis actually gives a lot of answers about what is working and what needs to be changed.

Chapter 9.

What We Can Do in Our Communities

English social reformer Octavia Hill is one of the least-known but most amazing people I've read about. She has one of the greatest legacies for the visual environment of Britain. Octavia did many amazing things in her life, from helping to take children off the streets of London and into education and work, to creating what is now the modern-day concept of social housing and spreading it around the world, but it was something she started in 1895 that will forever cement her in history for the impact she has had on the British people.

On 12 January, 1895, Octavia Hill co-founded the National Trust for Places of Historic Interest or Natural Beauty so that green spaces could 'be kept for the enjoyment, refreshment, and rest of those who have no country house . . . forever, for everyone'.[1] On that day, she acknowledged the right we all had to access nature and beautiful landscapes, whether we were rich or poor, by creating public parks. Today the National Trust protects thousands of rivers, coastlines, forests, parks, historic sites and monuments in the UK. Many of these would not exist today if it were not for this organisation protecting them and campaigning for them for over 125 years. Anyone who has ever enjoyed Hampstead Heath or Parliament Hill Fields in London has Octavia Hill to thank for that experience. She and many others campaigned to save open spaces from development and the National Trust has spread that ethos across the UK and beyond. Let's hope that more public art projects follow the same destiny.

In the Victorian era the only public art was open spaces and I certainly feel as passionate about the democratisation of those open spaces as Octavia did. We now have the opportunity to also include public art projects in our streetscapes to make them more engaging, useful and beautiful to further democratise our access to inspiring visuals.

In your visual audit, you certainly noted aspects and elements of the cityscape that had an effect on you. You decided whether or not you liked what those visuals presented. This is something that I do every day. While our cities are assuredly not displaying public art on every corner, let us consider now the effect art in your community's public spaces could have in changing your visual narrative.

As a cyclist, I find it fascinating that most of our streets have more space for the cars than for the cyclists and the pedestrians. Cars were invented in 1886 so our streets in their current form are less than 150 years old. What seems fixed and established can in fact change quite quickly. I hope that over the next hundred years, our streets can better integrate art as part of them, with less adverts and more space for pedestrians and cyclists so that we all benefit from a more positive visual landscape in our shared spaces. I have a strong instinct that as we start to realise the impact of the visual clutter we are exposed to, we will begin to take action. The good news is that most of us can get involved at a local level to support public art on our streets or in our parks.

What is Public Art?

Public art ranges from the exhibition of sculpture out-doors, to community murals, land art, site-specific art, performances and even street furniture. I have been passionate about public art for a long time, not only because it brings people together and enhances our visual consumption as we go about our busy lives but also because it can be functionally beneficial as well. And again, we also have a lot more say or influence around public art in our area than we might believe.

Why Public Art Matters

In the 1960s, there were only two megacities. Today we count over thirty megacities where we are bombarded by commercial imagery – and even in less dense urban areas its proliferation can be overwhelming. Public art introduces a visual rest in our overtly commercialised metropolises. It listens to us rather than imposing on us. It's integral to our quality of life, community self-image and identity. Cities are social spaces and artistic visuals introduce play, creativity and imagination into them. If you want to view art in the traditional sense you have to go to a museum or an art gallery. As such, the location we associate with art is an aside to life. A place that we visit. It's separated from life, and so museums and galleries can often feel a bit sterile or dead. I prefer art in life. Where the visual of public art accompanies us as we journey through our day. It is the art we see on the way to the train, or the art we see as we commute to work or the art we see when we are going for a walk or out shopping or walking our dog. It exists in real time. It's vibrant and energised and forever changing and evolving as we change and evolve.

There has been a marked rise in public art projects in major cities around the world including London, where I live. On 3 June, 2017, outside London Bridge station, a terrorist vehicle-ramming and stabbing took place.[2] It was horrific, and as a result, councils started to erect metal bollards outside many London stations to prevent vehicles from driving into them. While this is a brilliant health and safety answer, these bollards were a stark reminder that we couldn't trust everyone in the city we lived in and we stood divided amongst ourselves. I decided to pitch an artistic answer to Network Rail (the main railway company across the United Kingdom) and London Bridge station where we would wrap each bollard with the art of our artist Jennifer Abessira so that the seventy-two bollards became an inspiring visual narrative rather than architectural objects that fuel fear and division. I surveyed over 800 people on the ground when it was up and the perception change of the area, thanks to the public art project, was unanimous. Post a traumatic event, we had been able to write a new narrative for the area.

Another very personal and meaningful public art example was during the Covid lockdown in London. In the United Kingdom, we were allowed to walk around, as long we had no physical contact with another human being. I remember walking around and seeing all the shops closed, and that feeling of a huge ghost town. Nurses, doctors, cleaners, truck drivers, etc. were still hard at work. We pitched to our existing clients Westminster Council and The Crown Estate to start displaying art in all the shop windows; we called it Silver Lining. As an artist agency, we get – and at that time, got – feedback about the effect of the art we display on people's mindset and their experience of the city. We get texts, emails and private messages on Instagram almost every single day on the public art projects that we do, and these messages often start by saying, 'I was waiting for my bus, my tube, or I was just walking quickly with no objective whatsoever to encounter anything cultural or artistic. And then I stopped when I saw it.' Many of the messages we receive are from people who've never been to a museum or an art gallery. They are removed from the cultural and creative scene so public art allows us to give them a taste of it and creates a moment for them to pause, contemplate and absorb a meaningful visual message that isn't trying to sell them something, changing their experience of the city for a moment from an often overwhelming and commercial visual landscape to a place of peace, reflection and inspiration.

In March 2017, I spoke about the future vision we had of smart cities at the Royal Society of Arts. My hypothesis was – and remains – that while contemporary research into future cities tends to focus on technology, architecture and infrastructure, I see the installation of visual art projects as a cost-effective response to social inequalities created by population growth and stronger political division. One of my inspirations to think this way was former New York Mayor Michael Bloomberg who wielded public art as a powerful tool, one integral to building the 'confidence' of citizens, and invested a total of $2.8 billion in cultural organisations.[3] He transformed Times Square by adding art to the iconic displays to create moments of peace between advertisements, and hired a permanent art director for the city to oversee public art projects thereafter, setting a trend for big cities everywhere to view public art as a valuable source of revenue. One memorable example of how public art can help a city's bottom line was Christo and Jeanne-Claude's project, The Gates, in Central Park, which drew 4 million visitors who, in turn, brought an estimated $254 million to the tourism industry in 2005.[4]

Public art can stop us in our tracks, give us a moment to contemplate, a beautiful place to eat our lunch or enjoy a coffee. It is part of life and is living alongside us. It can also help boost our cities financially and unite us with messages that bring us together. Now a lot of the examples I have given you were made possible because they were backed by larger organisations, but they demonstrate how relatively small changes can have a powerful impact.

Public Art as a Tool for Change

Many of the problems we face globally and in our local communities require an artist's mind and local engagement. Everything from income inequality, unemployment, poverty, education and healthcare, through to isolation and loneliness, can be alleviated with creativity and critical and visual thinking. Artists can be truth tellers, offering transcendental experience in an increasingly literal world. Art can challenge us to feel something new, to connect to our identity and our common humanity. The people in our communities, you, me and our neighbours, also have ideas and creative input about how to make our communities safer, more appealing, how to represent our shared and evolving heritage, and how to create more shared space. And yet the two rarely meet.

Every community has artists, every community has people who want to get involved in making their surroundings better for everyone. Increasingly, local decision makers are tapping into this rich resource. Not only to better understand community needs and place making in general but to bring these ideas to fruition. Whether that be a gathering place, a gateway, wayfinding elements such as extra lighting in art as we did in Tower Hamlets, decorative fences, a mural, street furniture (benches, tree grates, tree guards, bollards), pathways, sculpture, sculptural streetlights, façade improvements, a fountain, earthworks, pavement decorations – the options are endless. Over the past three years in the United Kingdom, most of the public art briefs my public art director Serra Ataman received from councils and real estate developers asked her to specifically look for local artists and include the people who lived and worked there to vote for the artist that they wanted. This is an excellent example of how artists should be at the centre of society. That is all we advocate as a firm: that creatives and artists should be actors of change. It means that we should hear their voices more, by being a direct part of the conversation.

In 2022, Laura Zabel, executive director of Springboard for the Arts, which operates Creative Exchange, a platform for sharing free toolkits and resources for artists and communities, published a piece in the *Guardian* about an exciting movement operating around the world to share ideas and models that help connect artists more deeply with their communities.[5] From Santiago, Chile, to St Paul, Minnesota, USA, local citizens are partnering with artists to address challenges and make positive change. She wrote about several projects, including a Community Supported Art movement in the US and Canada that approaches art like community-supported architecture by creating communal ways to fund the creation of new artworks. Other projects, like the Gap Filler project in New Zealand that creates temporary installations and community spaces out of vacant properties, and the Irrigate project in the US which mobilises hundreds of artists to turn neighbourhoods disrupted by major infrastructure projects into destinations with murals and performances that recontextualise the construction zone, focusing on using art to combat the negative visual impacts of public spaces being abandoned, damaged or remodelled. These projects and thousands like them all over the world demonstrate what's possible when decision makers, artists and the community come together to create more inspiring space and solve real problems in communities. Public art can achieve great things – enhance gathering places, make a positive first impression which attracts residents and businesses. It also brings in visitors who spend time in the area, improving wayfinding and safety in the area when the art is also functional and serves a needed additional purpose.[6] Public art also allows the community to personalise a neighbourhood and build character and shared identity. It allows the locals to tell their stories or create new ones. Altering the visual consumption of residents, it also visually relays the character of a neighbourhood, city or town: a way to demonstrate neighbourhood pride and stitch the fabric of that community together against vandalism and neglect. Public art is a demonstration of community care and a desire to invest in collective quality of life. By transforming community eyesores into community assets, graffiti and murals, or major infrastructure like substations or water treatment plants into beautiful structures that contribute to the district instead of detracting from it, we improve everyone's visual consumption as they move around their local area. Public art is also an amazing way to build bridges between people who may not ordinarily meet and that can only be a positive outcome as we move forward in a multicultural, diverse world.

The Influence of Business Improvement Councils

In the UK, public art largely falls under the remit of local councils and Business Improvement Districts (BID).[7] Councils collect and use wider taxpayer money to provide services and, potentially, public art projects, and BIDs are using money collected from businesses in the area to improve the environment. Depending on tax revenue some councils have entire art departments which are hugely tuned in to how beneficial public art is for the community – and others don't.

I am constantly working with councils and BIDs and what is evident is a genuine desire, regardless of budget and budgetary constraints, to engage with the local community. But the word is not always getting out. They may ask for input on their website or they may broadcast on their social media feeds, but unless the audience is reading the council website or has followed their council on social media – the audience, in this case, residents in the local community – they just don't know about the consultation process. Or they know and are cynical about whether their input will be listened to. In my experience, hardly anyone from the local community turns up for these sessions, even when they are on Zoom. Instead, there is either no one from the community in these community engagement sessions or the same three people turn up. Often, they don't even care about public art, they just want a platform to air some other type of grievance! The challenge starts here: your life is busy and you rarely come into contact with your council. Help us make public art projects democratic by vocalising to your council that you wish to take part in the voting process, following their social media and listening to the poor chap who is giving you a track about it on the streets. There is nothing more heartbreaking than knowing that people have these rights and yet they do not exercise them. The streets could tell our visual stories, but only if we all wish to contribute to this narrative.

Our democratic access and engagement with public art can and will only change when the size of the participating audience increases. These institutions are using taxpayer money so we all have a very legitimate right to be part of the decision-making process.

Zero participation creates a triple-loss scenario. The council or BID loses because they don't get the ideas and input from the community about what they want and need. They assume the lack of engagement means that people don't care so they either carry on without community engagement and waste money on things that the community doesn't want and then complains about, or they redirect that money elsewhere so the community feels that the councils don't care. Either way the councils, BIDs, residents, local businesses and community all lose out.

The opposite is also true. If there is a large audience of interested and engaged community participants, ideally representative of the community, then they can give ideas and feedback to the council and BIDs. That way, not only do the councils and BIDs get to say that they consulted with, say, 1,000 people in the community, they also have confidence that the initiative is wanted and needed. Because the local community is involved in the development of the idea of that initiative, they are brought into it and those in the initial discussion panel can act as advocates for the project in the community to foster even greater buy-in and support. And of course, the wider community then benefits from the improvements in place making.

The councils and BIDs need to get better at communicating their need for input and the community needs to get better at making sure they are included in the discussions. There is no perfect systemic process yet, so for now simply Google your local area and BID or contact your local council and tell them that you would love to get more involved in local issues regarding place making, regeneration, development and public art – ask who you need to speak to or what list you need to get on to be kept informed.

Councils are generally quite responsive and we all know we can complain to the council and someone will get back to us, so there is nothing to stop us reaching out to request that we are put on a public consultation list. The more people act, the more the process will be built around it.

Remember, the only way to democratise public art and make our community spaces places for us all to enjoy is for more of us in the community to get involved in the discussions and decision making. And it is our right to be able to do so. If this doesn't change then art, including public art, will stay unequal and undemocratic.

Most councils and BID groups will be delighted to hear from you – so get in touch. The two work hand in hand so if you can't find your local BID online then your council will know how to contact them. If you have an idea of a new public art project in your neighbourhood, speak to your local representative – they can tell you about your local processes for proposing new projects.

Chapter 10.

Re-imagining Your Cities and Neighbourhoods

Just as we can vote with our likes and shares online to support visual messaging and creators we believe in, increasing their visibility not only within our own digital landscape but also in the digital landscapes of others, we also have a vote when it comes to our outside spaces and public art.

There are three main ways in which we cast our vote. The first is purely and simply the choice of where we live. The places you choose to frequent, live in and spend time in show your community that this is the sort of visual environment you prefer. That's one way we can vote. The second way is to start or join conversations about changing the visuals around you. Politicians will strive to implement the changes that we demand most and repeatedly. A case for public art must be made strongly. The third way is to take matters into your own hands and get involved in public art projects directly, which is what this section is all about. Public art is the visual storytelling we wish to see in our everyday, it needs to tell as much about us as we would wish for others. It needs to be our safe space within an overwhelming city.

In your visual audit, you will have noted aspects and elements of the cityscape or your shared spaces. You decided you liked what those visuals presented or you did not. Since our towns and cities are assuredly not displaying public art on every corner, there are plenty of opportunities for you to make your mark, so consider what changes could be made in your community to improve the spaces around you. These can be small spaces like the entrance to your local school, the shared spaces in your apartment building, your local park or even your street. Once you identify what you would change, you can start to take steps to turn your vision into a reality.

Your personal involvement, as we will see, is key. With the knowledge and awareness all these preceding pages have provided for you, consider that if we shape the conversation around visuals, they become ours. The visual world belongs to all of us. We all get to craft it and we should all get to participate in it and the choices behind it; it should be democratic.

Whether you have decided that you want to transform your town centre or start smaller, you can follow these steps to put together a plan and make your vision a reality. The same strategies apply to changing a visual in your town, apartment building, school or HOA – the main difference between these shared and community spaces is who the stakeholders are and the scale of the project.

Your Community/City Visual Detox Plan: How to Get Involved

If you want to enact change, the next steps are to define a need as you see it, decide what aspect of the community visual narrative you will focus on, select a site, determine a budget (of time, energy or money you'll invest) and select an artist or build a team to accomplish the project. You can do that alone, or in a group or network or collective of like-minded individuals.

1. Define a Need

Maybe your neighbourhood needs to acknowledge and better understand its history or the challenges faced by certain members of your community. This was recently the case in Cape Town and in Washington DC where two projects were used to help unite the community. I loved the public art project 'Beyond Walls' by our French artist Saype, painted in Langa township, Cape Town, in January 2021.[1] This project was conducted with 350 locals who helped in applying the paint and defining its narrative. The history of Langa is very heavy as this is where Black South Africans were forced to move in 1923 after the Urban Areas Act under apartheid.[2] It's amazing to be able to give this place a new spotlight and narrative after this tragic history. The other strong recent example happened in Washington DC, where the plaza was renamed by Mayor Muriel Bowser on 5 June, 2020, after the Department of Public Works painted the words 'Black Lives Matter' in 35-foot (11-metre) yellow capital letters as part of the George Floyd protests.[3] For me, both are art history in the making, not in museums but on the streets instead.

When transforming a shared space, it is important to bring in other people who share that space to get them to support your project. You can start this process by requesting written feedback, creating a discussion group on social media or gathering everyone together for a real-time conversation, either in person or online. Tell the group about the problem you've identified and invite people to make suggestions about what they would like to address. Seek feedback from all types of relevant stakeholders, businesses and residents, young and old. You could organise a walk-through and invite people along to really assess your visual environment and what could be improved in the very space you want to change.

There are many different ways that public art can be used to improve your community including:

- **Sense of place**: The creation of unique community places either by defining or redefining public spaces. This can help to give identity to the larger community as well as celebrate smaller areas. Art can become a focal point, beautiful as well as functional. The chosen art can also tell the communities' stories and translate those stories in unique visual forms.

- **Shared infrastructure**: In time, and with our democratic encouragement at local and national levels, public art will, I hope, become an integral part of urban planning and design. Combining functionality with art allows potentially ugly, bland structures to be transformed into joyous visual experiences and meaningful public symbols. On a smaller scale, this can be as simple as redecorating a break room.

- **Cultural infrastructure**: Are there any public art opportunities in your area that could highlight the cultural and historical connections within your community through local history, environmental systems, diverse cultural traditions and visual symbols?

- **Landmarks**: Are there any landmarks that the community would like to remove or repurpose? The sculpture of slave owner Edward Colston is an example of this replacement, albeit through protest.[4] Or could new landmarks be created to serve as beacons, build community pride and reinforce community identity? Two London sculptors, Thomas J. Price and Veronica Ryan, created in 2022 the first permanent UK public artworks in honour of the Windrush generation – Caribbean people who came to London between 1948 and 1970.[5]
- **Public buildings**: Could old buildings benefit from some external art to enhance civic pride and engagement or could art be included in the creation of new buildings?
- **Temporary works**: Could art bring joy to temporary works and the disruption they sometimes create, similar to the Irrigate project? Could that art draw more people to the area and become a short-term celebration for the community? The impermanence of these types of projects can be a good testing ground as they are never going to become permanent art pieces, so getting permission and funding may be easier. On a small scale, this can be as simple as hanging up

paintings in the hallway for a few weeks or putting up seasonal decorations.

2. Select a Site and Define Parameters

Where will the art make the most impact to the greatest number of people in your community? The art should transform or enhance the area. The site should be somewhere that provides the best canvas for the art, not too big, not too small for the area selected.

If you can get input and create guidelines for an artist to work within, even better. Aim for enough direction to ensure the community gets what you want but broad enough to give the artist room for creativity.

3. Determine the Budget and Identify Funding Sources

As you define your project and priorities, do research to determine what it would cost to bring it to reality. All of the artists, craftspeople and tradesmen involved will need to be paid fairly for their time. Get quotes from potential collaborators and experts in the field. Once you know how much money you would need, you can explore opportunities for funding. Don't be intimidated if the cost sounds out of reach for you personally – there are lots of opportunities to get funding for community projects. The simplest way is to split the cost with other members of your community who share your vision or put together a fundraiser, but you can also find funding opportunities through different organisations around you.

Here are some ways to identify sources of funding for larger community projects:

- Research local art organisations and become a member. Engage with the conversations that impact your local community. Follow the relevant organisations on social media. They can be a good resource not only for connecting you to artists but also for information on expected costs, etc.
- Check out the Art Fund, Arts Council or Art UK.
- Public Art Online has a useful page of links.
- Contact your local council or BID.

Ideally, for larger projects the best strategy is to get financial support from the local council or BID but there are also options within the local community, fundraising days, sponsorships from local businesses or donation of materials, and arts foundations may offer grants. Volunteers can help prepare the site, help the artist with installation and save time and money in many other ways. If you are leading it from your business internally, pitch it to the business as an initiative you would like your firm to support.

4. Select an Artist

Some projects will require you to bring in experts to help create a new visual element for the space you are transforming. There are several ways to find an artist or artists for your project. This is similar to finding the right candidate for a job. You can either hire an artist directly, or request submissions from a variety of artists. This is also true for any contractors and craftsmen you might need to realise the vision.

Are there any artists already living in your community that would like to get involved? Their connection to the area and the fact that they are a neighbour will almost always make the art even more special. Many of the artists on our books at MTArt Agency also specialise in public art. You can check out some of the portfolios at mtart.agency.

You Can Do It!

Today, half the world's population resides in urban areas. By 2050, two-thirds of us will be living in cities.[6] What will make our cities more liveable? All experts align: nature and art, places where we can gather, converse and literally breathe with less air, noise and visual pollution. Our visual consumptions are more and more impacted by the visuals that exist in urban spaces – coping with high-density visuals is not a future issue, but a current one that we can and should deal with.

A highly dense visual landscape that is overly targeting us, that is affecting (for better or worse) our health and well-being on all levels is our present experience. It will intensify if we don't take some action to mitigate our visual narrative. I do feel this effort goes beyond just community. It's also a collective conversation, a collective reimagining, a collective new vision of the future of what the cities can look like to be our safe landing place, our comforting home, our productive work environment.

When you're envisioning how to make all of your spaces – everywhere you live, work, travel and see things and people interacting – more environmentally friendly, more liveable, more visually attractive, more sensorily desirable, know that you can have a direct impact on the spaces that you share with others. It all starts with having conversations about what you see.

Conclusion

Visual Literacy and the Future of Visuals

Claire Luxton, one of our artists at MTArt Agency, created a powerful image in 2020 that went viral. It was her own face with a tear running down the left cheek that led to a small blossoming bunch of flowers under her left eye. Tears are very symbolic in mythology as often they lead to a more positive turn of the story once the hero/heroine has shown vulnerability and been through the hardest of times. It felt like spring after winter. It was, in fact, the first of her images that I ever saw before we started working together. When I looked at that image, I thought, 'That's a visual of Hope. Hope for happiness and joy after a very difficult time.' You will see her image through your own lens, of course. But for me, I couldn't help but feel that the pain we were collectively experiencing during the pandemic, over a longer term, would really be healed. The subliminal but easily detected message of this piece is that ultimately a circumstance, a situation, an issue, or a problem we are facing is difficult at the moment, but it is possible for that painful thing to blossom into something else. A solution, a positive shift, a transformation awaits us all – it's the idea we can start afresh.

While we can easily become caught up in our supercharged visual world, consuming visuals at an ever-increasing rate to the point where we lose touch with the effect they are having on us, it is never too late to take some time for a visual detox and start again with a new, intentional and mindful, visual consumption habit.

While my awareness of the visuals around me is certainly amplified because I work in the art world, we all have the ability to be more in tune with the visual messages we are receiving and more intentional about which ones we consume. I feel a deep sense of responsibility to help people recognise that we are all surrounded by art and beauty in our everyday life: the beauty in nature, the beauty in the faces of those we love and the beauty of new ideas and human ingenuity and creativity. What we see on a daily basis is our art, it is our gallery and it has a profound impact on who we believe we are, what we believe is possible for us and how we feel day to day. We need to stop thinking that art is something else – over there, not for us. We all live within a shared visual landscape that is our collective art. It's time to become much more aware of our visual consumption so we can curate the visual narrative we want in our own lives.

In this book, I've ultimately given you the tools and the vocabulary not only to analyse your visual environment but to discuss visuals with others. Maybe up until reading this book, you could only verbalise, 'I like it,' or 'I don't like it,' about what you saw. You'd walk into a space or out onto the street and just get a good feeling or an unpleasant one. Now, though, with this book, I'd like you to be able to say, 'I can speak that visual language, I know where it came from and why it was created, I can interact with it, I can contribute to it, I can participate in it.' With these skills, you can not only improve your life and wellbeing, but also help the people around you to better understand their visual consumption and confidently take action within your community to make our shared visual landscape empowering and inspiring for all of us. You'll be able to see through how governments, unscrupulous media outlets and individuals, as well as corporations, are trying to manipulate you through what you see and decide for yourself what you believe and how you want to see yourself. My hope is that you will now take up the baton and help us all to create a new, less money-driven, less elite, and more democratic way to approach the visual culture. We can create a visual narrative where everyone gets to see themselves represented on billboards, on walls, in homes – and where we recognise a more inclusive, democratic and inspiring world. We can have a world where visuals once again become a core language used to bring us together.

Your 4-Week Visual Detox Plan

This detox plan is meant to be a challenge – it may feel a bit uncomfortable at times, but by sticking to the whole plan and reflecting on it daily it will help you discover how changing your visual landscape affects you emotionally, and you may discover new habits and practices that you want to keep in your life going forward.

Week 1 – Build a New Habit

This week, the challenge is to start habits that will reduce your digital consumption every day for the whole detox. Do each of these things every day and continue to do them daily throughout the detox:

1. Do not check your phone or look at a screen for the first hour of your day. If your phone is also your wake-up alarm, use an alarm clock or install a screentime limiting app on your phone that helps you resist the urge to check your phone after turning off your alarm. For example, Apple's built-in Screen Time app has features to help you limit your access to your phone.
2. Do not check your phone or look at a screen for the last hour of your day. Instead, take ten

minutes to reflect on how your visual detox is making you feel and prioritise other activities, such as tidying up clutter, laying out your clothes for the next day, reading or meditating.

3. Start logging the number of hours you spend in different kinds of spaces, including online, in nature and in commercial spaces, such as shopping centres.

Week 2 – Reduce Your Digital Consumption

In addition to continuing the digital detox from last week, the goal for this week is to take a few minutes every day to remove visuals from your visual landscape.

Day 1:

1. Go through your email inbox and unsubscribe from as many newsletters as you can in twenty minutes.
2. Take a photo of your bedroom, then look around and remove one visual that is not bringing the energy you want for that space. This might mean moving your laundry basket to the closet or cleaning your nightstand. Take a photo when you're done so that you can reflect on how the change makes you feel.

Day 2:

1. Go through your phone and delete any apps that you are not using. You can always re-download any you find you need again later, so be ruthless and delete as many as you can.
2. Take a photo of your kitchen, then look around and remove one visual that is not bringing the energy you want for that space. Take a photo after you're done and reflect on whether you feel differently looking at the space afterwards.

Day 3:

1. Disable notifications on as many apps as you can (only keep notifications where they are absolutely necessary).
2. Take a photo of your workspace, then take fifteen to twenty minutes to tidy it up. Are there any items that you have stored or organised in a new way? Are there any items that distract you that you can store out of sight? Take a photo when you're done so that you can compare the two and reflect on how the change makes you feel.

Day 4:

1. Unfollow or mute any accounts on your feed that are not providing consistently interesting and inspiring content.
2. Choose one cupboard in your house and take a photo of it. Then, clear it out, removing every item, cleaning the inside of the cupboard, and only putting back the items that are still good. Take a photo at the end and compare how you feel looking at each one.

Day 5:

1. Install an adblocker on all of your browsers on your internet-capable devices.
2. Have a no-buy day. Avoid all unnecessary purchases. If you're going to be out of the house during meal times, pack a snack or lunch to avoid having to buy food or drinks while you're out. Keep track of all of the moments that you feel an urge to buy something and make a note of which visual sparked that desire.

Day 6:

1. Set a timer for when you're consuming intense visuals today, such as watching a movie or the television or looking at social media, then spend the

same amount of time deliberately focusing on 'slow' visuals, such as reading print media, looking out the window, taking a walk through your neighbourhood or mindfully doing household chores without any distractions immediately afterwards.

2. Go through one section of your wardrobe, such as your sock drawer or your shirts, and remove any pieces that you aren't wearing regularly. You can choose to put these pieces into storage for now. As you move through the rest of the detox, reflect on how it feels to have more space to see all of your belongings and to only see the pieces that you actually need.

Day 7:

1. Have a zero-screentime day: what is the minimum amount of time you can spend looking at any screens today? This includes television, films, answering emails and social media.
2. Deliberately avoid commercial visuals all day. Plan your routes to avoid areas with a lot of outdoor advertising and avoid commercial spaces such as shopping centres.

Week 3 – Quiet the Visual Noise

1. Put your phone on grayscale mode for the whole week. If you need the colour function for work,

turn it on during your working hours before turning it off again when you finish work for the day.

2. Spend twenty minutes outside each day without any technology. Seek out green spaces if you can. This can involve going to a local park or nature spot but could also mean just walking around your neighbourhood. Dress appropriately for the weather and spend the time looking for the beauty in the natural world – can you see any wildlife? What kind of plants do you notice? What was the sky like each day?

Week 4 – Seek Out New Visuals

For this week, seek out new visuals each day by doing each of these things:

1. Try a new route: whether you're running errands or commuting, change your route or mode of transportation. Can you find a more visually pleasing commute, even if it takes a few more minutes? Is there a route that minimises the amount of advertisements you see along the way?

2. Spend five minutes with art. You can go to an art gallery or museum or borrow an art book from the library, go to a local art exhibit or public art installation, or visit a museum's gallery online – however you find it, choose one piece of art, set a timer for five minutes, and spend the entire time

only looking at the artwork. Think about how it makes you feel and what it is trying to communicate with you visually.

3. Add one new visual that sparks joy or makes you feel at peace with your visual landscape. This can be a new background on your phone, a new piece of art on your walls, a new photo on your fridge, or simply reorganising your bookshelf by colour.

After the detox

Once you've finished your 4-week detox, go back through the daily reflections you made on how changing your visual consumption made you feel. Which parts of the detox has had the most positive impact for you? Consider which new habits you would like to keep moving forward.

Acknowledgements

When this book was commissioned by Square Peg I was pregnant with my second son Vivaldi and undergoing a violent cyberbullying campaign consisting of two hundred death threats sent to me daily via various social media platforms. Emma Beddington, a journalist at the *Guardian*, wrote in November 2022 that that year had been the worst on record for digital hate campaigns against female leaders, resulting in many of them stepping down. As it stands, as a female founder I have a 90% reduced chance of raising funds compared to a male founder but a 120% greater chance of experiencing online hate.

So this book took a village.

I am beyond lucky to work daily with an incredible team at MTArt Agency. Two of my juniors, Gabriel Shelsky and Bibiana Battistoni, rose to the challenge of this book and researched every single visual study under the sun. No one within the agency vocalised that perhaps it was madness to undertake this publishing deal and instead I received support from all my team, artists, partners and investors.

A huge thank you to my artists: what a dream you have been, stretching, challenging and expanding my visual brain on a daily basis.

To my investors, you understood from the start that the bold vision we had required a full rethink of the way in which we interact daily with our visual world.

Acknowledgements

To my super team, you embraced a leader who was seeking both economic and intellectual answers to the many challenges we uncovered in the visual sector.

To Sarah Kuehl, my dear friend from Warwick University, and Tamma Ford, thank you for finding words when I couldn't.

This book is a dream come true, and to be trusted with such an endeavour has humbled me, big time. I revere the publishing world and specifically Penguin Random House for striving to make complex and necessary topics engaging to all audiences. I wanted this exact bedrock for *The Visual Detox* so that we could introduce a new way of thinking to so many of you. I want to specifically thank Gail Rebuck, Grace Gould and my wonderful agent Gordon Wise for making it happen. The Square Peg team has been a dream to work with and my editor Marianne Tatepo challenged me daily, all the while reassuring me that I was capable of doing it, especially when I was pumped with postpartum hormones. Emily Martin and Graeme Hall, you have made my life a breeze.

I read recently in the *Harvard Review* that the media and the general public use very different words when describing male and female leaders. Rarely do we call a female leader 'smart', so I feel immensely fortunate that many of my primary and secondary teachers, and the most incredible business people whom I have looked up to over the years, have made me feel that my way of seeing the world is valid. It gave me an immense confidence. My first boss, Steve Lazarides, was one of them, alongside my very first art collector Yannick Pons, Michael Ovitz during my time in Los

Acknowledgements

Angeles and, most recently, one of our investors, Todd Ruppert. I wouldn't have been capable of writing this book and feeling legitimate in doing so without the many people who gave me this voice. So, whoever needs to read this today, I hope you can find that same confidence to follow your own voice.

As always, it is what you do outside of your profession that makes the difference. It's the smiles of my children, the hugs of my friends and the many times I danced with my ballet teacher Giovanna Lamboglia that made it possible to come back to the same text and tweak it again and again until it was right to present it to you, the reader. I followed my own advice and went to search for the answers in many different visual environments, while speaking to so many visual thinkers. I loved it. I ran back home many times, dancing to 'Modern Love' by David Bowie with my brain exploding with the excitement and pure joy of witnessing my ideas deepen.

I hope *The Visual Detox* gives you as much joy as it gave me when writing it, and that you carry it everywhere with you, tear out its pages and highlight the parts that connect directly with your life. It's entirely yours now.

Endnotes

While care has been taken to ensure that the web links in this book are accurate at the time of publication, the publisher cannot guarantee that these links remain viable.

Introduction

1 Paula Dootson, '3.2 billion images and 720,000 hours of video are shared online daily. Can you sort real from fake?', *QUT*, www.qut.edu.au/study/business/insights/3.2-billion-images-and-720000-hours-of-video-are-shared-online-daily.-can-you-sort-real-from-fake

2 T.J. McCue, 'Why Infographics Rule', *Forbes*, 8 January, 2018

3 Carine Alexis, '29 Incredible Stats that Prove the Power of Visual Marketing', *Movable Ink*: movableink.com/blog/29-incredible-stats-that-prove-the-power-of-visual-marketing

4 Amy Wakeham, 'Meet Marine Tanguy, The Entrepreneur Taking The Art World By Storm', *Country & Townhouse*, www.countryandtownhouse.com/culture/meet-marine-tanguy-the-entrepreneur-taking-the-art-world-by-storm/

1. Understanding Your Visual Consumption

1 Blake Ross, 'Aphantasia: How It Feels To Be Blind In Your Mind', Facebook, www.facebook.com

2 Joel Pearson, Thomas Naselaris, Emily A. Holmes, and Stephen M. Kosslyn, 'Mental Imagery: Functional Mechanisms and Clinical Applications', *Trends in Cognitive Sciences* 19, no. 10 (2015), pp. 590–602.

3 Soyiba Jawed, Hafeez Ullah Amin, Aamir Saeed Malik and Ibrahima Faye, 'Classification of Visual and Non-visual Learners Using Encephalographic Alpha and Gamma Activities', *Frontiers in Behavioural Neuroscience* 13, no. 86 (2017).

4 A. J. Drenth, 'Cognitive Styles of Thinkers (T) vs. Feelers (F): Visual, Spatial & Verbal', *Personality Junkie*, personalityjunkie. com/04/cognitive-styles-thinkers-feelers-visual-spatial-verbal/

5 Temple Grandin, *Visual Thinking: The Hidden Gifts of People Who Think in Pictures, Patterns and Abstractions* (Rider, 2022)

6 Temple Grandin, *Visual Thinking*

7 stemettes.org/about/, *Stemettes* (Accessed 28/15/23)

8 Bessel van der Kolk, *The Body Keeps the Score: Brain, Mind, and Body in the Healing of Trauma* (Penguin, 2015)

9 Jia Tolentino, 'The Age of Instagram Face', *New Yorker*, 12 December, 2019

10 Rob Binns, 'Netflix Statistics 2023: Subscriber amount, time watched, and platform growth', *Independent,* updated 3 August, 2023

2. We Live in a Very Visual World

1 'Nicotine Dependence', *Camh*, www.camh.ca/en/health-info/mental-illness-and-addiction-index/nicotine-dependence

Endnotes

2 Richard Gunderman, 'The Manipulation of the American Mind: Edward Bernays and the birth of public relations', *The Conversation,* 9 July, 2015

3 Suzanne Halliburton, 'John Wayne Once Revealed His Only Regret from His Career in 1976 Fan Interview', *Outsider,* outsider.com/entertainment/john-wayne-once-revealed-only-regret-career-1976-fan-interview/; Tom Poster and Phil Roura, 'Bette Davis' Final Tally Adds Up to Paltry Sum', *Los Angeles Times,* 6 April, 1990

4 Alexi Duggins, 'Why The Social Dilemma is the Most Important Documentary of our Times', *Independent,* 18 September, 2020

5 'Social Media and Mental Health', *HelpGuide,* www.helpguide. org/articles/mental-health/social-media-and-mental-health. htm#:~:text=A%202018%20University%20of%20Pennsyl vania,%2C%20sleep%20problems%2C%20and%20FOMO

6 Columbia University Medical Center, 'This Is Your Brain On Violent Media', *ScienceDaily,* 2007, www.sciencedaily.com/ releases/2007/12/071206093014.htm

7 David Galiana, 'The 5W1H Method: Project Management defined and applied', *Wimi,* www.wimi-teamwork.com/blog/ the-5w1h-method-project-management-defined-and-applied/ #:~:text=Definition,by%20analysing%20all%20the% 20aspects

3. Understanding Your Visual World

1 Manfred Zimmermann, 'Neurophysiology of Sensory Systems', *Fundamentals of Sensory Physiology,* ed. Robert F. Schmidt (Springer-Verlag Berlin, 1986)

Endnotes

2 John Coates, *The Hour Between Dog and Wolf: Risk-Taking, Gut Feelings and the Biology of Boom and Bust* (Fourth Estate, 2012)

3 David Foster Wallace, *This Is Water: Some Thoughts, Delivered on a Significant Occasion, about Living a Compassionate Life* (Little, Brown and Company, 1st Edition, 2009)

4 Daniel J. Simons and Christopher F. Chabris, 'Gorillas in our Midst: sustained inattentional blindness for dynamic events', *Perception* 28, no. 9 (1999)

5 Claire Harrison, 'Visual Social Semiotics: Understanding How Still Images Make Meaning', *Technical Communication* 50, no. 1 (2003), pp. 46–60. .

6 Yuval Noah Harari, *Sapiens* (Jonathan Cape, 2020)

7 'World Population Review: Hong Kong Population 2023 (Live)', *World Population Review*, worldpopulationreview.com/countries/hong-kong-population

8 James W. Bisley and Michael E. Goldberg, 'Attention, Intention, and Priority in the Parietal Lobe', *Annual Review of Neuroscience* 33 (2010), pp. 1–21

9 Amna Rehman and Yasir Al Khalili, *Neuroanatomy, Occipital Lobe* (StartPearls Publishing, 2023)

10 Chloe Bennett, 'What is the Neocortex', *News Medical & Life Sciences*, www.news-medical.net/health/What-is-the-Neocortex.aspx

11 Jon H. Kaas, 'Neocortex', *Encyclopedia of the Human Brain,* ed. Vilayanur S. Ramachandran (Academic Press, 2002)

12 Geri Stengel, 'How To Prevent The Pandemic From Taking a Greater Toll On Women Entrepreneurs', *Forbes*, 3 April, 2020

13 'School teacher workforce', *Gov.uk*, www.ethnicity-facts-figures.service.gov.uk/workforce-and-business/workforce-diversity/school-teacher-workforce/latest

14 Cady Lang, 'This Nike Commercial Features Women Athletes in Hijabs', *Time*, 24 February, 2017

15 Mike Savage, *Social Class in the 21st Century* (Pelican Books, 2015)

16 'National Statistics on the Creative Industries 2022', Creative Industries Policy and Evidence Centre, pec.ac.uk/news/national-statistics-on-the-creative-industries

17 Judith Duportail, *L'amour sous algorithme* (LGF, 2020)

18 E. I. Medvedskaya, 'Features of the Attention Span in Adult Internet Users', *Digital Society as a Cultural and Historical Context of Personality Development* 19, no. 2 (2022), pp. 304–319.

19 Marcel Proust, *In Search of Lost Time* (CreateSpace Independent Publishing Platform, 2015)

20 Betty Plewes and Rieky Stuar, *The Pornography of Poverty: A Cautionary Fundraising Tale* (Cambridge University Press, 2009)

21 Sandra Blakeslee, 'Using Rats to Trace Anatomy of Fear, Biology of Emotion', *New York Times*, 5 November, 1996

22 Olivia Guy Evans, 'What is the Limbic System? Definition, Parts and Functions', *Simply Psychology*, updated 13 October, 2023

23 Kerry J. Ressler, 'Amygdala Activity, Fear, and Anxiety: Modulation by Stress', *Biological Psychiatry* 15, no. 67 (2010) pp. 1117–19

24 Amit Goldenberg and Robb Willer, 'Amplification of Emotion on Social Media', *Nature Human Behaviour* 7, no. 6 (2023), pp. 845–46

25 M. J. Crockett, 'Moral Outrage in the Digital Age', *Nature Human Behaviour* 1, (2017) pp. 769–71

26 Dan Vergano, 'Rescue of Ancient Ruin of Pompeii Follows New Plan', *National Geographic*, 19 April, 2014

27 Albert Read, *The Imagination Muscle: Where good ideas come from (and how to have more of them)* (Constable, 2023)

28 Branwen Jeffreys, 'Creative subjects being squeezed, schools tell BBC', BBC, www.bbc.co.uk/news/education-42862996

29 Hickmore, Harry, 'Cuts Mean Arts Education Is Being Outsourced To The Culture Sector – And It's Not Working', *Huffington Post*, 16 April, 2019

30 Scott Reyburn and Anny Shaw, 'Private sector picks up the pieces as UK government cuts art education funding', *The Art Newspaper*, 19 September, 2023

31 Francis Green and David Kynaston, *Engines of Privilege* (Bloomsbury, 2019)

32 Orian Brook, Dave O'Brien and Mark Taylor, *Culture is Bad for You: Inequality in the Cultural and Creative Industries* (Manchester University Press, 2020)

33 Mark Brown, 'Tate and Steve McQueen call for "arts-rich" school curriculum', *Guardian*, 17 December, 2019

34 Max Roser, *Talent is everywhere, opportunity is not. We are all losing out because of this* (OurWorldInData, 2019)

35 Stacy Liberatore, 'Mathematician reveals how YOU can increase your chances in winning the lottery – and why you should NOT have 'lucky numbers', Mailonline, 5 April, 2023

4. The Anatomy of a Visual Message

1 C. Harrison, 'Visual Social Semiotics: Understanding how still images make meaning', *Technical Communication, ProQuest Central* 50, no. 1, 2003, p. 46

2 John Lichfield, 'The moving of the Mona Lisa', *Independent*, 5 April, 2005

Endnotes

3 Morriss-Kay, Gilliam M. 'The Evolution of Human Artistic Creativity', *Journal of Anatomy and Physiology*, 216, no. 2 (2016), pp.158–76.

4 Anna Rahmanan, 'We Don't All Wear Black At Funerals. Here's What Mourners Wear Across Cultures', *Huffington Post*, www.huffingtonpost.co.uk/entry/wearing-black-at-funerals_l_640a12fee4b0653e296a4d35

5 T. Chen, P. Samaranayake, X. Cen, M. Qi and Y.-C. Lan, 'The Impact of Online Reviews on Consumers' Purchasing Decisions: Evidence From an Eye-Tracking Study', *Frontiers in Psychology* 13 (2002), pp. 1–13

6 Lizzy Hillier, 'How brands use colour psychology to reinforce their identities', *Econsultancy*, econsultancy.com/how-brands-use-colour-psychology-to-reinforce-their-identities/

7 Anna Gruener, 'Vincent van Gogh's yellow', *British Journal of General Practitioners* 63, no. 612 (2013) pp. 370–71

8 Gruener, 'Vincent van Gogh's yellow'

9 Jon Henley, 'Why Vote Leave's £350m weekly EU cost claim is wrong', *Guardian*, 10 June, 2016

10 'Colour Symbolism in Chinese Culture: What Do the Traditional Colours Mean?', *Colour Meanings*, www.colour-meanings.com/colour-symbolism-in-chinese-culture-what-do-traditional-chinese-colours-mean/

11 'The Artist is Present', *MomaLearning*, www.moma.org/learn/moma_learning/marina-abramovic-marina-abramovic-the-artist-is-present-2010/#:~:text=Over%20the%20course%20of%20nearly,with%20me%2C%E2%80%9D%20Abramovi%C4%87%20explained

12 Kenneth Brummel, *Picasso: Painting the Blue Period* (DelMonico Books, 2022)

13 'Dove | Reverse Selfie | Have #TheSelfieTalk', YouTube, www.youtube.com/watch?v=z2T-Rh838GA

14 Michel Pastoureau, *Green: The History of a Colour* (Princeton University Press, 2014)

15 Ayami Oki-Siekierczak, '"How green!": The Meanings of Green in Early Modern England and in *The Tempest*' (Laboratoire d'Etudes et de Recherches sur le Monde Anglophone, 2015)

16 'Why Does Copper Turn Green', *Metal Supermarkets*, www.metalsupermarkets.co.uk/why-copper-turns-green/#:~:text=Copper%20oxide%20can%20in%20some,further%20corrosion%20and%20increases%20durability

17 Samantha David, 'Where life imitates art: France's literary love affair with Molière', *The Connexion*, www.connexionfrance.com/article/People/Profiles/Where-life-imitates-art-France-s-literary-love-affair-with-Moliere

18 Olivia B. Waxman, 'How Green Became Associated With St. Patrick's Day and All Things Irish', *Time*, 16 March, 2016

19 Robert Jobson, 'The Queen's coronavirus speech illuminated on Piccadilly Circus billboard as UK lockdown continues', *Evening Standard*, 8 April, 2020

20 Sali Hughes, *Our Rainbow Queen: a Celebration of our Beloved and Longest-Reigning Monarch* (Square Peg, 2019)

21 Christopher Hope, 'Ukip's yellow and purple colours and pound sign logo could be dropped as it reinvents itself in "post-Brexit age"', *Telegraph*, 27 March, 2017

22 Kendra Cherry, 'What Does the Color Purple Mean? Purple Color Meaning and Psychology', *Very Well Mind*, www.verywellmind.com/the-colour-psychology-of-purple-2795820

Endnotes

23 '50 facts about Queen Elizabeth II's Coronation', (fact 39), *Royal*, www.royal.uk/50-facts-about-queen-elizabeth-iis-coronation

24 Rosa Silverman and Ben Wright, 'The "pink collar" crisis strangling the wellbeing – and economy – of Britain', *Telegraph*, 29 April, 2023

25 Chloe Malle, 'Imperial Pink? The Wing Gears Up to Go Global', *Vogue*, 8 August, 2018

26 Sara Ashley O'Brien, 'The Wing, the once-buzzy women's coworking community, has closed its doors', CNN, edition.cnn.com/2022/08/31/tech/the-wing-startup-closes/index.html

27 Wang, Jenna, 'The Wing's Audrey Gelman On Building A Feminist Co-Working Empire', *Forbes,* 13 August, 2018

28 Anna North and Chavie Lieber, 'The big, controversial business of The Wing, explained', *Vox*, 7 February, 2019

29 Amanda Hess, 'The Wing is Woman's Utopia. Unless You Work There', *New York Times*, 17 March 2020

30 Agencies, 'Barbie movie tops $1bn in global box office ticket sales, breaking record for female directors', *Guardian,* 7 August, 2023

31 Claire Wilcox, *Fashioning Masculinities: The Art of Menswear* (V&A, First Edition, 2022)

32 Lauren Cahn, ' How Pink and Blue Became the "Girl" and "Boy" Baby Colours', *Readers Digest*, updated 13 November, 2022

33 'Pink Colour Meaning: The Colour Pink Symbolises Love and Compassion', *Colour Meanings*, www.colour-meanings.com/pink-colour-meaning-the-colour-pink/

34 Kendra Cherry, 'The Meaning of the Color Brown in Psychology', *Very Well Mind*, www.verywellmind.com/the-colour-psychology-of-brown-2795816#:~:text=Colour%20Brown%20Meaning%20in%20Feng%20Shui&text=2%20Brown%20represents%20either%20wood,lack%20of%20ambition%20and%20drive

35 'Colours of Mourning Around the World', *Funeral Guide*, www.funeralguide.co.uk/blog/mourning-colours

36 'Are brown eggs healthier than white eggs?', *Healthline*, www.healthline.com/nutrition/white-vs-brown-eggs#:~:text=Often%2C%20people%20who%20prefer%20brown,white%20eggs%20are%20healthy%20foods

37 'The History of Earth Day', www.earthday.org/history/

38 Kendra Cherry, 'The Meaning of the Color Brown in Psychology'

39 Simran, 'Colour on Branding Business', *SGJ Co.*, simran-jaiswal.com/colour-on-branding-business

40 Yasmina Reza, *Art* (Faber & Faber, 1996)

41 Tracey Wallace, 'The History of Mourning Dress and Attire in the West', *Eterneva*, www.eterneva.com/resources/mourning-dress

42 'The Victoria Connection', *V&A*, www.vam.ac.uk/articles/the-victoria-connection#:~:text=Queen%20Victoria%20is%20commonly%20credited,her%20role%20as%20the%20monarch

43 Alyssa Newcomb, 'Why this *Vogue Italia* cover was left blank intentionally', *Today*, 10 April, 2020

44 'Origin of Batman', *Wikipedia*, en.wikipedia.org/wiki/Origin_of_Batman

45 'Most Popular Car Colours in America', *GermainCars*, www. germaincars.com/most-popular-car-colours/#:~:text= Regardless%20of%20gender%2C%20grayscale%20colours, vehicles%20on%20the%20road%20today

46 Charity Claypool, 'The Surprisingly Simple Reason Your Next Car Should Be Silver', www.motorbiscuit.com/the-surprisingly-simple-reason-your-next-car-should-be-silver/

47 'Versailles' golden gates recreated after 200 years', *France 24*, www.france24.com/en/20080702-versailles-golden-gates-recreated-after-200-years-france-monuments

48 A. J. de Craen, P. J. Roos, A. L. de Vries and J. Kleijnen, 'Effect of colour of drugs: systematic review of perceived effect of drugs and of their effectiveness', *BMJ* 28, no. 313 (1996) p. 313

49 Lynda Mead, *The Female Nude: Art, Obscenity and Sexuality* (Routledge, 1992)

50 John Berger, *Ways of Seeing* (Penguin Modern Classics, 2008)

51 Martha M. Lauzen, 'It's a Man's (Celluloid) World: Portrayals of Female Characters in the Top Grossing U.S. Films of 2022', San Diego State University, womenintvfilm.sdsu.edu/wp-content/uploads/2023/03/2022-it's-a-man's-celluloid-world-report-rev.pdf

52 Nina Myashita, 'You've Heard Of The Madonna-Whore Complex, But Do You Really Know What It Means?', *Refinery29*, www.refinery29.com/en-gb/what-is-madonna-whore-complex

53 Lauren Michele Jackson, 'The Invention of "the Male Gaze"', *New Yorker*, 14 July, 2023

54 Asare, Janice Gassam, 'Understanding The White Gaze And How It Impacts Your Workplace', *Forbes,* 28 December, 2021

55 Heidi Mitchell, 'How a Model's Eye Gaze Can Make an Ad More Effective', *Wall Street Journal*, 19 February, 2021

5. Context Matters

1 Jane Wakefield, 'TED 2018: Fake Obama video creator defends invention', BBC, www.bbc.co.uk/news/technology-43639704
2 Masha Gessen, 'The Photo Book That Captured How the Soviet Regime Made the Truth Disappear', *New Yorker*, 15 July, 2018
3 Erin McCann, 'Civil war veterans at Gettysburg anniversary in 1913 – in pictures', *Guardian*, 1 July, 2013
4 Katie Rogers, 'Protestors Dispersed With Tear Gas So Trump Could Pose at Church', *New York Times*, 1 June, 2020
5 Lamide Akintobi, '"Westernization is not the answer": Artist Àsìkò explores Yoruba culture through mythology', CNN, edition.cnn.com/style/article/asiko-ade-okelarin-artist-nigeria-spc-intl/index.html
6 Melanie Schefft, 'Why you may need a social media scheduler', University of Cincinnati, www.uc.edu/news/articles/2023/03/uc-cid-shares-why-you-may-need-a-social-media-scheduler.html
7 Stokel-Walker, Chris, 'We Spoke to the Guy Who Created The Viral AI Image of the Pope That Fooled the World', *Buzzfeed News*
8 Grace Shao, 'Social media has become a battleground in Hong Kong's protests', CNBC, www.cnbc.com/2019/08/16/social-media-has-become-a-battleground-in-hong-kongs-protests.html

9 Sam Gregory, 'Deepfakes, misinformation and disinformation and authenticity infrastructure responses: Impacts on frontline witnessing, distant witnessing, and civic journalism', *Sage Journals* 23, no. 3 (2021)

10 Shelly Banjo, Natalie Lung, Annie Lee and Hannah Dormido, 'How Hong Kong's Leaderless Protest Army Gets Things Done', *Bloomberg UK*, www.bloomberg.com/graphics/2019-hong-kong-airport-protests/

11 Jim Brunner, 'Fox News runs digitally altered images in coverage of Seattle's protests, Capitol Hill Autonomous Zone', *Seattle Times*, 12 June, 2020

12 Natalie Jomini Stroud, 'Polarisation and Partisan Selective Exposure', *Journal of Communication* 60, no. 3 (2010), pp. 556–76

13 Pennycook, Gordon and David G. Rand, 'The Psychology of Fake News', *Trends in Cognitive Sciences* 25, no. 1 (2021)

14 Eli Pariser, *The Filter Bubble: What The Internet Is Hiding From You* (Penguin, 2012)

15 Dan Milmo, 'Social media firms "monetising misery", says Molly Russell's father after inquest', *Guardian*, 30 September, 2022

16 Azi Paybarah and Brent Lewis, 'Stunning Images as a Mob Storms the U.S. Capitol', *New York Times*, 6 January, 2021

17 Zach Schonfeld, 'Conservative group finds "absolutely no evidence of widespread fraud" in 2020 election', *The Hill*, thehill.com/homenews/campaign/3559758-conservative-group-finds-absolutely-no-evidence-of-widespread-fraud-in-2020-election/

18 Shelly Grabe, L. Monique Ward, Janet Shibley Hyde, 'The role of the media in body image concerns among women: a

meta-analysis of experimental and correlational studies', *Psychological Bulletin* 134, no. 3 (2008), pp. 470–76.

19 John Naughton, 'The growth of internet porn tells us more about ourselves than technology', *Guardian*, 30 December, 2018

20 Luke Hurst, 'Generative AI fueling spread of deepfake pornography across the internet', www.euronews.com/next/2023/10/20/generative-ai-fueling-spread-of-deepfake-pornography-across-the-internet

21 Monica Corcoran Harel, 'Do Fillers Actually Make You Look Older?', *Elle*, 22 July, 2016

22 Kiamani Wilson, 'Cosmetic Surgery Consumption and the Influence of Celebrity Prominence', Haverford College Thesis, 2018

23 Candice E. Walker, Effects of social media use on desire for cosmetic surgery among young women, *Current Psychology* 40 (2021)

24 *ASPS National Clearinghouse of Plastic Surgery Procedural Statistics* (American Society of Plastic Surgeons, 2020)

25 Marika Tiggemann, '#Loveyourbody: The effect of body positive Instagram captions on women's body image', *Elsevier* 33 (2020), pp. 129–36

26 Daniel Arkin, '"Gorgeous piece of public art": World's largest self-anchored suspension bridge opens in San Fran', NBC News, www.nbcnews.com/news/us-news/gorgeous-piece-public-art-worlds-largest-self-anchored-suspension-bridge-flna8c11071013

27 'Model Legislation for Banning Billboards', *Scenic America*, www.scenic.org/take-action/resources/model-legislation-for-banning-billboards/

28 'This is the Teenage Brain on Social Media', *Neuroscience*, neurosciencenews.com/nucleus-accumbens-social-media-4348/

29 Aviv Weinstein, 'Neurobiological mechanisms underlying internet gaming disorder', *Dialogues Clinical Neuroscience* 22, no. 2 (2020), pp. 113–26

30 Kylie Jenner, Instagram www.instagram.com/kyliejenner/?hl=en; Louvre, Instagram www.instagram.com/museelouvre/?hl=en

31 Judith Duportail, Nicolas Kayser-Bril, Edouard Richard, Kira Schacht, 'The skin bias in Instagram', *Voxeurop_*, voxeurop.eu/en/the-skin-bias-in-instagram/

32 Josh Taylor, 'TikTok received more requests to remove child bullying posts than any other social platform in Australia', *Guardian*, 20 July, 2023

33 Anisha Kohli, '"We Can Turn It Off." Why TikTok's New Teen Time Limit May Not Do Much', *Time*, https://time.com/6259863/tiktok-time-limit-teens/

7. Transforming Your Digital Landscape

1 Laura Silver, 'Smartphone Ownership Is Growing Rapidly Around the World, but Not Always Equally', Pew Research Centre, www.pewresearch.org/global/2019/02/05/smartphone-ownership-is-growing-rapidly-around-the-world-but-not-always-equally/#:~:text=For%20example%2C%20a%20median%20of,country%2C%20even%20across%20advanced%20economies

2 'Ad Blocking Is Here to Stay and Can No Longer Be Ignored', *Adtoniq*, www.adtoniq.io/blog/ad-blocking-here-to-stay#:~:text=Globally%2C%2047%25%20of%20internet%20users,over%20one%20billion%20devices%20worldwide; 'Consumer

attitudes towards digital advertising and ad blocking usage', *Insider Intelligence*, www.insiderintelligence.com/insights/ad-blocking/

3   Sarah Finley, 'You can now block "dangerous and triggering" weight loss ads from appearing on Instagram', *Women's Health*, updated 18 August, 2022; Omar Oakes, 'Brands accused of funding terror groups through online ads', *Campaign Live*, www.campaignlive.co.uk/article/brands-accused-funding-terror-groups-online-ads/1423717

4   Neelfyn, 'How to Make the Most of AdBlock', *GetAdBlock*, blog.getadblock.com/how-to-make-the-most-of-adblock-304f0e9dc1bc

5   Chris van Tulleken, *Ultra-Processed People: Why Do We All Eat Stuff that Isn't Food . . . and Why Can't We Stop?* (Cornerstone Press, 2023)

6   Elizabeth Bernstein, 'Three Findings That Changed the Way We Think About Sex', *Wall Street Journal*, https://www.wsj.com/articles/what-kinseys-75-years-of-sex-research-says-about-us-11662464272

7   Will Green, 'Campaign Best Places to Work 2023', *Campaign Live,* https://www.campaignlive.co.uk/article/revealed-campaign-best-places-work-2023/1817327

8   'Community Labels', *Tumblr*, help.tumblr.com/hc/en-us/articles/5436241401239-Community-Labels

9   'The future of moderation is participatory', *Reliabl*, reliabl.ai/

10 Jesselyn Cook, 'Women Are Pretending To Be Men On Instagram To Avoid Sexist Censorship', *Huffington Post*, www.huffingtonpost.co.uk/entry/women-are-pretending-to-be-men-on-instagram-to-avoid-sexist-censorship_n_5dd30f2be4b0263fbc99421e

11 Eli Erlick, 'How Instagram May Be Unwittingly Censoring the Queer Community', *Them*, www.them.us/story/instagram-may-be-unwittingly-censoring-the-queer-community; 'Bridging the Gap: Local voices in content moderation', *Article 19*, www.article19.org/bridging-the-gap-local-voices-in-content-moderation/; Kalhan Rosenblatt, 'Months after TikTok apologised to Black creators, many say little has changed', NBC News, www.nbcnews.com/pop-culture/pop-culture-news/months-after-tiktok-apologized-black-creators-many-say-little-has-n1256726; Julia Angwin, 'Facebook's Secret Censorship Rules Protect White Men From Hate Speech But Not Black Children', *ProPublica*, www.propublica.org/article/facebook-hate-speech-censorship-internal-documents-algorithms; Todd Spangler, '40% of LGBTQ Adults Do Not Feel Safe on Social Media, GLAAD Survey Finds', *Variety*, 13 July, 2022; Aaron Sankin, 'How activists of colour lose battles against Facebook's moderator army', *Reveal*, revealnews.org/article/how-activists-of-colour-lose-battles-against-facebooks-moderator-army/

12 Robyn Conti, 'What Is An NFT? Non-Fungible Tokens Explained', *Forbes Advisor*, www.forbes.com/uk/advisor/investing/cryptocurrency/nft-non-fungible-token/

13 Jex Exmundo, 'Quantum: The Story Behind the World's First NFT', *NFT Now*, nftnow.com/art/quantum-the-first-piece-of-nft-art-ever-created/

14 Eileen Kinsella, 'Sotheby's Nets $17 Million With Its First-Ever NFT Auction (Which Included Almost 20,000 Very Fungible Works)', *ArtNet News*, news.artnet.com/market/sothebys-first-ever-sale-of-nfts-pak-and-nifty-gateway-1959276

15 William Entriken, Dieter Shirley, Jacob Evans and Nastassia Sachs, 'ERC-721: Non-Fungible Token Standard', *Ethereum Improvement Proposals* (2018)

16 Jacob Kastrenakes, 'Beeple sold an NFT for $69 million through a first-of-its-kind auction at Christie's', *The Verge*, www.theverge.com/2021/3/11/22325054/beeple-christies-nft-sale-cost-everydays-69-million

17 Daniel Spielberger, 'Art and NFTs: Beeple reflects one year after historic $69 million digital art sale', NBC News, www.nbcnews.com/tech/tech-news/art-nfts-beeple-reflects-one-year-historic-69-million-digital-art-sale-rcna18989

18 'The Art Basel and UBS Global Art Market Report', Art Basel, www.artbasel.com/stories/art-market-report-2021?lang=en

19 Neal Stephenson, *Snow Crash* (Penguin, 1994)

20 Daniel Kahneman, *Thinking Fast and Slow* (Allen Lane, First Edition, 2011)

21 Matthew A. Killingsworth, Daniel Kahneman, and Barbara Mellers, 'Income and emotional well-being: A conflict resolved', *University of Virginia* 120, no. 10 (2022)

8. Reimaging Your Visual Landscape

1 'Lindsay T. Graham, S. Gosling and Chrisopher K. Travis, 'The Psychology of Home Environments', *Perspectives on Psychological Science* (1 May, 2015)

2 'Viewing Art Gives Same Pleasure as Being in Love', *UCL*, www.ucl.ac.uk/news/headlines/2011/may/viewing-art-gives-same-pleasure-being-love

3 Gabriel A. Orenstein and Lindsay Lewis, *Erikson's Stages of Psychosocial Development* (StatPearls Publishing LLC, 2022)

4 Alan C. K. Cheung and Robert E. Slavin, 'How Methodological Features Affect Effect Sizes in Education', *Sage Journals* 45, no. 5 (2016)

5 'Blue Light has a Dark Side', *Harvard Health Publishing*, Harvard Medical School, 7 July, 2020

6 Marie Kondo, *The Life-Changing Magic of Tidying Up: The Japanese Art of Decluttering and Organizing* (Ten Speed Press, 2014)

7 Jura Koncius, 'Marie Kondo's life is messier now – and she's fine with it', *Washington Post*, 26 January, 2023

8 Michelle Ma, 'Dose of nature at home could help with mental health, well-being during Covid-19' *UW News*. 17 April, 2020

9 Albert Read, *The Imagination Muscle* (Constable, 2023)

9. *What We Can Do in Our Communities*

1 'National Trust celebrates 125 years', *Discover Britain*, discoverbritainmag.telegraph.co.uk/national-trust-125/#:~:text=It%20has%20been%20a%20remarkable,take%20curious%20and%20tremendous%20forms%E2%80%9D

2 'London Bridge attack: What happened', BBC, www.bbc.co.uk/news/uk-england-london-40147164

3 Marine Tanguy and Vishal Kumar, 'Measuring the extent to which Londoners are willing to pay for public art in their city', *Science Direct*, www.sciencedirect.com/science/article/abs/pii/S0040162518318249

4 'Christo's Gates: Big $ for Big Apple Artist's central park project brought in $254 million in economic activity for the city:

mayor', *CNN Money*, money.cnn.com/2005/03/03/news/newsmakers/gates/

5 Laura Zabel, 'Six creative ways artists can improve communities', *Guardian*, 12 February, 2015

6 Melissa G. Bublitz, 'Collaborative Art: A Transformational Force within Communities', *Chicago Journals* 4, No. 4

7 'Business Improvement Districts: The Role of BIDs in London Regeneration', *London Assembly*, 2016

10. Reimagining Your Cities and Neighbourhoods

1 'French artist SAYPE paints massive mural at foot of Mount Royal', *Global News*, globalnews.ca/video/9747790/french-artist-saype-paints-massive-mural-at-foot-of-mount-royal

2 Sophia Hyatt, 'South Africa's housing crisis: A remnant of apartheid', *Aljazeera*, www.aljazeera.com/features/2016/10/11/south-africas-housing-crisis-a-remnant-of-apartheid

3 Lauren Aratani, 'Washington mayor stands up to Trump and unveils Black Lives Matter mural', *Guardian*, 6 June, 2020

4 Archie Bland, 'Edward Colston statue replaced by sculpture of Black Lives Matter protester Jen Reid', *Guardian*, 15 July, 2020

5 Amah-Rose Abrams, '"It's not a monument, it's a celebration": Windrush sculpture unveiled in Hackney', *Guardian,* 22 June, 2022

6 '68% of the world population projected to live in urban areas by 2050, says UN', United Nations Department of Economic and Social Affairs, https://www.un.org/development/desa/en/news/population/2018-revision-of-world-urbanization-prospects.html

About the Author

Marine Tanguy is the CEO and founder of the MTArt Agency, a talent agency based in London. MTArt Agency was the first talent agency in the art world and is now the leading one globally, disrupting the way in which artists are represented and democratising a traditionally elitist sector. She managed her first art gallery aged twenty-one and then co-founded a gallery in Los Angeles at twenty-three. Tanguy was listed in *Forbes*'s 2018 '30 Under 30' and was named NatWest's 2019 'UK Entrepreneur of the Year'. MTArt was listed as one of the 100 fastest-growing companies in the UK by *The Times* in 2022. She is a writer and keynote speaker on contemporary art, art investment and how art can be used to shape our lives. Her talks include two TEDx Talks on how to transform cities with art and how social media visuals affect our minds.